Electronics
Circuits and Systems

Fourth Edition

Owen Bishop

ELSEVIER

AMSTERDAM • BOSTON • HEIDELBERG • LONDON • NEW YORK • OXFORD • PARIS
SAN DIEGO • SAN FRANCISCO • SINGAPORE • SYDNEY • TOKYO

Newnes is an imprint of Elsevier

Newnes

Newnes is an imprint of Elsevier
The Boulevard, Langford Lane, Oxford OX5 1GB, UK
30 Corporate Drive, Burlington, MA 01801

First edition 1999
Reprinted 2000
Second edition 2003
Third edition 2007
Fourth edition 2011

British Library Cataloguing-in-Publication Data
A catalogue record for this book is available from the British Library

Library of Congress Cataloging-in-Publication Data
A catalog record of this book is available from the Library of Congress

ISBN: 978-0-08-096634-2

For information on all Newnes publications visit
our website at www.newnespress.com

Printed and bound in Italy

11 12 13   10 9 8 7 6 5

Working together to grow
libraries in developing countries

www.elsevier.com | www.bookaid.org | www.sabre.org

ELSEVIER   BOOK AID International   Sabre Foundation

Contents

This book is written for a wide range of pre-degree courses in electronics. The contents have been carefully matched to current UK syllabuses at Level 3 / A-level, but the topics covered, depth of coverage, and student activities have been designed so that the resulting book will be a student-focused text suitable for the majority of courses at pre-degree level around the world. The only prior knowledge assumed is basic maths and the equivalent of GCSE Double Award Science.

The UK courses covered by this text are:

BTEC National Engineering Pathways syllabus (2007), Units 5 (Electrical and Electronic Principles), 35 (Principles and Applications of Electronic Devices and Circuits), 62 (Microprocessor Systems and Applications), and the introductory stages of Units 51 (Industrial Process Controllers), 60 (Principles and Applications of Analogue Electronics), 68 (Principles and Applications of Microcontrollers), and 90 (Telecommunications Principles).

A-level (AS and A2) specifications from AQA, OCR and WJEC.

The book is essentially practical in its approach, encouraging students to assemble and test real circuits in the laboratory. In response to the requirements of certain syllabuses, the book shows how circuit behaviour may be studied with a computer, using circuit simulator software.

The book is suitable for class use, and also for self-instruction. The main text is backed up by boxed-off discussions and summaries, which the student may read or ignore, as appropriate. There are frequent 'Self Test' questions at the side of the text with answers given in Supplement B. Numeric answers to other questions and answers to the multiple choice questions are on the companion website.

The text has undergone a major revision to produce this fourth edition. Additions to the content include five new Topics. These cover electrical and magnetic fields, diodes, oscillators, integrated circuits, and industrial process control systems. Several other Topics have been expanded, to reflect the increasing importance of digital electronics and microcontroller systems. All Topics have been updated where necessary, to keep pace with the many recent developments in electronics.

The 'On the Web' panels in many Topics are a new feature of this edition. They are intended for students to make use of the wealth of relevant information available from that source. Also, this edition coincides with the launching of a companion website. This has a PowerPoint presentation of illustrations from the book for use by students and lecturers. It has the answers to numeric questions and to all the multiple choice questions. There are many more of these in this edition. There are pages of worked examples and questions for those who need extra support in maths.

The companion website also includes some novel features: a set of calculators for electronic formulae, animated diagrams to show electronic circuits in action, and series of interactive worksheets, with answers.

Owen Bishop

PRACTICAL CIRCUITS AND SYSTEMS

Circuit Ideas

As well as being a textbook, this is a sourcebook of circuit ideas for laboratory work and as the basis of practical electronic projects.

All circuits in this book have been tested on the workbench or on computer, using a circuit simulator. Almost all circuit diagrams are complete with component values, so the student will have no difficulty in building circuits that will work.

Testing Circuits

The circuit diagrams in this book provide full information about the types of components used and their values. Try to assemble as many as you can of these circuits and get them working. Check that they behave in the same ways as described in the text. Try altering some of the values slightly, predict what should happen, and then test the circuit to check that it does.

There are two ways of building a test circuit:

- Use a breadboarding system to build the circuit temporarily from individual components or circuit modules.
- Use a computer to run a circuit simulator. 'Build' the circuit on the simulator, save it as a file, and then run tests on it.

The simulator technique is usually quicker and cheaper than breadboarding. It is easier to modify the circuit, and quicker to run the tests and to plot results. There is no danger of accidentally burning out components.

CONVENTIONS USED IN THIS BOOK

Units are printed in roman type: V, A, s, S, μF.
Values are printed in italic (sloping) type:

Fixed values	V_{CC}, R_1
Varying values	v_{GS}, g_m, i_D
Small *changes* in values	v_{gs}, i_d

Resistors are numbered, R1, R2, and so on. The *resistance* of a resistor R1 is represented by the symbol R_1. The same applies to capacitors (C1, C2) and inductors (L1, L2).

Significant Figures

When working the numerical problems in this book, give the answers to three significant figures unless otherwise indicated.

Units in Calculations

Usually the units being used in a calculation are obvious but, where they are not so obvious, they are stated in square brackets. Sometimes we show one unit divided by or multiplied by another.

Example

On p. 72, we state:
  $R_1 = 14.3/2.63$ [V/μA] $= 5.44$ MΩ

A voltage measured in volts is being divided by a current measured in *micro*amperes.

Mathematically, this equation should be written:
  $R_1 = 14.3/(2.63 \times 10^{-6}) = 5.44 \times 10^6$

Set out in this form, the equation is difficult to understand and to remember. To avoid this problem we quote the *units* instead of powers of 10. When the result is being worked out on a calculator, it is easy to key in the values (14.3, 2.63) and follow each by keypresses for 'EXP -6' or other exponents where required. The result, in Engineering or Scientific format, tells us its units. In this example the display shows 5.437262357^{06}. We round this to 3 significant figures, '5.44', and the '06' index informs us that the result is in megohms.

COMPANION WEBSITE

The URL of the site is: http://www.elsevierdirect.com/companions/9780080966342

Part 1

Circuits

An electrical circuit is a pathway for the flow of electric current.

In a **direct current** (or **DC**) circuit, the current flows in one direction. It flows from a point in the circuit that is at a high potential to a point in the circuit that is at a low potential. The circuit may be simple loop (like the one shown here) or it may be a network of two or more branches.

The circuit diagram (or schematic)

Explaining the diagram:

B1 is a battery of 2 cells, total emf = 3 volts (or 3 V).
S1 is a switch, single pole double throw (SPDT).
D1 is a light-emitting diode (LED).
R1 is a resistor, resistance = 56 ohms (or 56Ω)

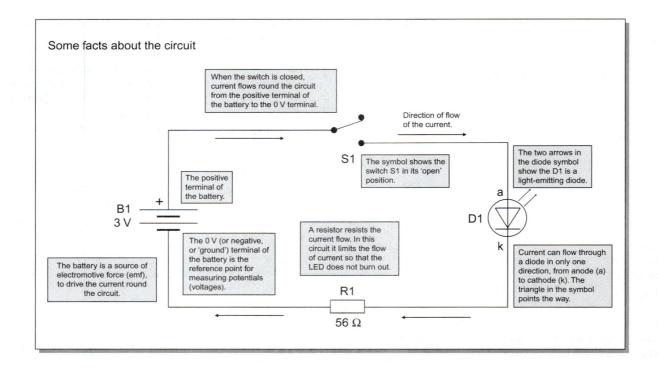

Some facts about the circuit

When the switch is closed, current flows round the circuit from the positive terminal of the battery to the 0 V terminal.

Direction of flow of the current.

The symbol shows the switch S1 in its 'open' position.

The two arrows in the diode symbol show the D1 is a light-emitting diode.

The positive terminal of the battery.

The 0 V (or negative, or 'ground') terminal of the battery is the reference point for measuring potentials (voltages).

A resistor resists the current flow. In this circuit it limits the flow of current so that the LED does not burn out.

Current can flow through a diode in only one direction, from anode (a) to cathode (k). The triangle in the symbol points the way.

The battery is a source of electromotive force (emf), to drive the current round the circuit.

A **conductor** is a material in which electric current can flow. All metals are conductors. List the conductors in this circuit. An **insulator** (or **non-conductor**) is a material in which current can not flow. List the insulators in this circuit (some answers in Supplement B, p. 361).

The diode D1 is made from a special kind of material called a **semiconductor**.

An electric current is a flow of electrically charged particles.

A current flows in a circuit if there is a potential difference (pd) between two points in the circuit. A pd may be produced in many different ways:

- chemical action: as in an electric cell.
- electricity generator: uses the energy of burning fuel (coal, oil, gas).
- nuclear power station: uses heat from nuclear reactions.
- solar cells: use energy from sunlight.
- wind farm: uses energy from wind.

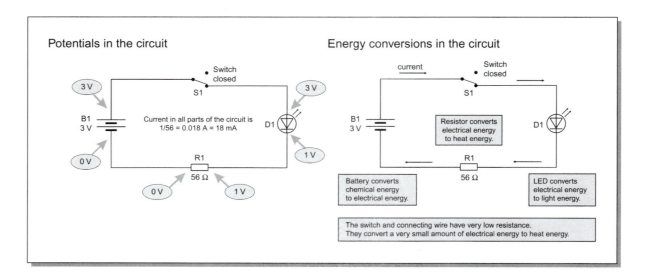

Potentials in the circuit

Current in all parts of the circuit is 1/56 = 0.018 A = 18 mA

Energy conversions in the circuit

Resistor converts electrical energy to heat energy.

Battery converts chemical energy to electrical energy.

LED converts electrical energy to light energy.

The switch and connecting wire have very low resistance. They convert a very small amount of electrical energy to heat energy.

All of these convert one form of energy into another form, electrical energy.

Each part of the circuit above has:

- a **current** flowing through it, measured in amperes (or **amps** for short).
- a **pd** across it, measured in **volts**.
- a **resistance** to the flow of current, measured in **ohms**.

These three quantities are related by this equation:

$$current = \frac{pd}{resistance}$$

The equation expresses Ohm's Law.

In the circuit on p. 1, the battery is the source of electrical energy. There is a pd between its terminals (see diagram on p. 2, bottom right), making current flow in the circuit. The electrical energy from the battery is converted by the LED and the resistor to other forms of energy (light and heat; see lower diagram on p. 2, bottom). The rate at which energy is being converted is measured by its power, where power is calculated using this equation:

$$Power = pd \times current$$

The unit of power is the watt, symbol W. There are more equations for power, p. 354.

Taking the 0 V (or 'negative') terminal of the battery as the reference point, the other terminal is at 3 V. The wire and switch are very good conductors so the potential at the anode of the LED (D1) is also 3 V. There is a **forward voltage drop** of about 2 V across a conducting LED so the potential at its cathode is about 1 V.

There is a drop of 1 V across the resistor, bringing the potential down to 0 V.

Starting at the 0 V terminal of the battery and moving clockwise round the circuit there is a rise of potential across the battery. The potential drops across D1 and R1.

The total potential (or voltage) drop in a circuit equals the potential rise across the source of emf.

Self Test

There are some questions about volts, amps, ohms and watts on the companion website.

Diodes

A diode is made from semiconducting materials. It has two terminals, called the **anode** and the **cathode**. Current can flow easily from anode to cathode.

FIGURE 1.1 Typical low-power diodes. The diode on the left is a signal diode and the one on the right is a Zener diode. On both diodes the black band at one end indicates the cathode terminal.

CURRENT AND VOLTAGE

When there is a voltage (a pd) across a resistor, the two quantities are related as in the equation pd/current = resistance (see opposite). Ohm's Law applies.

This is not true for diodes. The circuit at top right has a milliammeter (mA) to measure the current through the diode, and a voltmeter (V) to measure the voltage across it. The power source is a power supply unit (PSU) that produces a DC voltage, variable from 0 V up to, say, 10 V.

The voltage is set to different values and the current is measured for each voltage. A graph of the results (the V-I graph) would be a straight line for a resistor, but is curved for a diode.

The diode in the circuit above is connected so that current flows freely through it, from anode (a) to cathode (k). It is said to be **forward biased**.

FIGURE 1.2 A circuit for measuring the voltage-current relationship of a forward biased diode.

V and I

By convention, *V* is the symbol for the size of a voltage. *I* is the symbol for the size of a current. (p. 353)

The **V-I graph** obtained with this circuit is shown below. Its main feature is that no current flows

FIGURE 1.3 The V-I curve is an example of a **transfer characteristic**. It links the changes in the output of a component or circuit (the current through the diode) to the changes in the input (the forward bias voltage).

through the diode until the forward bias of more than about 0.7 V. As the voltage is increased above 0.7 V, the current increases, slowly at first then more rapidly.

REVERSE BIAS

If a diode is connected the other way round, so that its cathode is more positive than its anode, it is **reverse biased**.

Only a few nanoamps (1 nA $= 10^{-9}$ A $=$ 0.0000000001 A) flow through a reverse biased diode. With a signal diode or rectifier diode, the reverse voltage may be up to several hundred volts, depending on the type. But, if the voltage is too big, the diode breaks down and is destroyed.

An LED has a small breakdown voltage. Typically, a reverse bias of about 5 V will destroy it. Check that the LEDs are connected the right way round before switching on the power supply to a newly-built circuit.

LEDs may also break down when forward biased if the current through them is too big. The way to prevent this is described on p. 6.

When reverse biased, a Zener diode breaks down at a fixed voltage, its Zener voltage, V_Z (see top

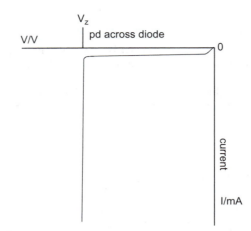

FIGURE 1.4 The v-i graph of a reverse biased Zener diode has a sharp 'knee' at the Zener voltage, v_z.

right). This may be only a few volts, Zener breakdown does not destroy the diode. This property is used in voltage regulator circuits.

Photodiodes are nearly always connected with reverse bias. The small leakage current through the photodiode is proportional to the amount of light falling on it.

EXTENSION BOX 1   Conduction in Semiconductors

An electric current in a metal is a flow of electrons (negative charge) from negative to positive. This is because the atoms of a metal have electrons that are able to break free and flow under the influence of an electric field.

Electron flow also occurs in a type of semiconductor material known as **n-type semiconductor**. This is a semiconductor such as silicon, to which has been added small amounts of an element such as antimony. The antimony provides additional electrons to increase the conductivity of the material. It is called n-type because electrons carry negative charge.

FIGURE 1.5 Doping pure silicon with antimony provides a free electron, which increases conductivity.

EXTENSION BOX 2 Conduction by Holes

Another type of semiconductor is known as **p-type semiconductor**. In this, the added material is an element such as indium. The indium atoms have a 'vacancy' in their outer electron orbit where an electron can be taken in. Such a vacancy is known as a **hole**.

The first stage is illustrated in (a).

At (1) an electron escapes from a silicon atom and drifts in the electric field until it finds a hole at the indium atom. At (2), the hole this created is filled by an electron escaping from an atom further along the material.

The next stage is illustrated in (b).

At (3), the hole created at (2) is filled by an electron from the negative supply, having travelled along a metal wire. At (4) an electron escapes and passes into the metal wire of the positive supply. The hole created is again filled at (5) by an electron, so shifting the hole toward the negative end.

The holes apparently move from positive to negative, in the opposite direction to the electrons. So we can think of holes as positive charge carriers. For this reason, silicon doped with indium or boron is known as **p-type silicon**.

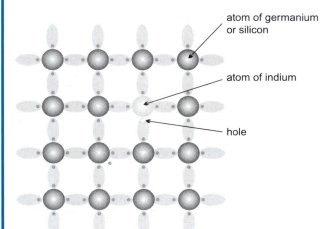

atom of germanium or silicon

atom of indium

hole

FIGURE 1.6

What happens when an electric field is present is shown below:

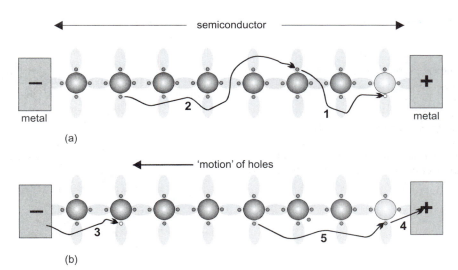

semiconductor

metal

2

1

metal

(a)

'motion' of holes

3

5

4

(b)

FIGURE 1.7 Conduction in **p-type** silicon. Holes move from positive to negative and so are equivalent to positive charge carriers.

EXTENSION BOX 3 Conduction in Diodes

Consider a bar of silicon, doped so that half is p-type and half is n-type. The bar is not connected to a circuit, so no external electric field is applied to it. Electrons and holes are free to wander at random.

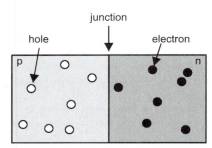

FIGURE 1.8

Some of the electrons from the n-type material wander across the junction, attracted by the holes in the p-type material. The holes in the region of the junction become filled. In the p-type region (to the left of the junction) the filling of holes means that the atoms have, on average, more electrons than they should. The atoms have become negative *ions*.

Similarly, in the n-type region near the junction the atoms have, on average, lost electrons leaving holes and so have become positive *ions*.

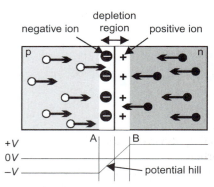

FIGURE 1.9

Although these ions are charged, they can not act as charge carriers. This is because they are *fixed* in their places in the lattice. The result is that the region on side A of the junction has overall negative charge and the region on side B has overall positive charge. There is a pd across the junction.

The effect of this pd is the same as if a cell was connected across the junction. This is not a real cell but we refer to its *effects* as a **virtual cell**. At a silicon p-n junction, the pd is about 0.7 V. At a germanium junction, it is about 0.2 V.

The region that contains ions has no mobile charge carriers and it is called the **depletion region**. We think of the pd across the depletion region as a **potential hill**, the potential rising as we pass from the p-type side to the n-type side. Eventually, as more carriers cross the junction, the hill becomes so steep that no more carriers can cross it.

If the semiconductor is connected into a circuit and a pd is applied to it, the junction can be **biased** in either direction. With **reverse bias**, the external source reinforces the action of the virtual cell, making the potential hill higher. This makes the depletion region wider than before, and there is even less chance of carriers finding their way across it. No current flows.

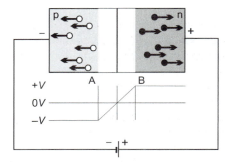

FIGURE 1.10

If the external source is connected so that its pd opposes that of the virtual cell, the junction is **forward biased**. The depletion region becomes narrower. The potential hill becomes lower. Electrons are more easily able to cross the junction.

If the external pd is greater than 0.7 V (for silicon), the virtual cell is overcome and charge carriers flow across the junction. There is a **forward voltage drop** of 0.7 V across the junction.

FIGURE 1.11

The p-n junction has the unusual property that it allows conduction in one direction (p to n) but not in the other. We find pn junctions in diodes, MOSFETs (p. 11) and in BJTs (p. 14). They show one-way conduction and a forward voltage drop of approximately 0.7 V when forward biased.

QUESTIONS ON CIRCUITS

1 Draw a schematic diagram of a door alert circuit built from a battery of 4 cells, a push-button, and a buzzer (see the chart of symbols on p. 357). On the diagram mark the point in the circuit that is at the highest potential. Show the direction of flow of current.

2 Draw a schematic diagram of a door alert circuit that has two push-buttons, A and B. The buzzer is to sound when either A or B is pressed. What happens if both buttons are pressed at the same time?

3 A DC electric motor is powered by a 4.5 V battery and controlled by a SPST (single pole single throw) switch. There is a light emitting diode which comes on when the power supply is switched on. The diode needs a resistor to limit the current through it. Draw the schematic diagram of the circuit.

4 List the energy conversions that occur when the circuit of Question 3 is switched on.

5 Name four good conductors of electric current. Give an example of the use of each and state which one of the four is the best conductor.

6 Name four electrical insulators, giving an example of the use of each.

7 The pd across a torch lamp is 2.4 V and the current through it is 700 mA. What is the resistance of the lamp?

8 What is the current through a 220 Ω resistor if the pd across it is 5 V?

9 What is the power of the torch lamp referred to in question 7?

10 What is the power of the resistor referred to in question 8?

11 A 2 V car battery has a 100 Ω resistor connected across its terminals. What is the current through the resistor? At what rate is electrical energy being converted and to what form is it converted?

On the Web

A lot of information about electronics can be found on the World Wide Web. Where you see a panel like this one in the book, access the Web on a computer and visit some of the listed sites. Use a browser such as *Internet Explorer, Safari, Netscape* or *Mozilla Firefox*.

For some topics, go to a search engine such as *Google* or *Yahoo!* and type in a keyword, such as 'diode'.

In *Yahoo!*, 'diode' gave us over 7 million references. These included tutorials and explanations such as those from *Wikipedia* and *How Stuff Works* sites, as well as manuacturers' data sheets and suppliers' catalogues.

Yahoo! has an *Answers* site (http://answers.yahoo.com), which covers many topics, including electronics (under Engineering). There you can post questions or try to answer those asked by others.

Search the Web for information on diodes and related topics. It may help you with some of the questions on this page and overleaf.

QUESTIONS ON DIODES

1 Name the terminals of a diode. In which direction does the current flow when the diode is forward biased?

2 Describe, with the help of a graph, how the current through a forward biased diode varies as the pd across the diode is varied from 0 V to 5 V.

3 Using a data sheet or information obtained on the Web, list and comment on the characteristics of a named signal diode.

4 Repeat 3 for a high-power rectifier diode.

5 List the features of six different types of LED, and state an application for each.

6 Draw the symbols for (a) a signal diode, (b) a Zener diode, (c) a photodiode, and (d) a light-emitting diode. Label the anode and cathode of each symbol.

7 Search this book, or another electronics book, or the WWW, for a simple electronic circuit containing a diode. Draw the circuit and describe how the properties of the diode are made use of in this circuit.

MULTIPLE CHOICE QUESTIONS

1 An example of a conductor is:
   A nylon.
   B aluminium.
   C rubber.
   D glass.

2 An example of an insulator is:
   A copper.
   B steel.
   C PVC (polyvinyl chloride).
   D carbon.

3 The pd across a resistor is 15 V and its resistance is 120 Ω. The current through the resistor is:
   A 125 mA.
   B 8 A.
   C 1.25 A.
   D 105 mA.

4 A silicon diode is forward biased so that its anode is 0.5 V positive of its cathode. The current through the diode is:
   A 0.5 A.
   B 0 A.
   C not known.
   D enough to burn out the diode.

Companion Site

A box like this tells you that the Companion Site for *Electronics – Circuits and Systems* has more about this topic.

The URL of the site is: http://www.elsevierdirect.com/companions/9780080966342. The site has questions on electrical circuits and Ohm's Law. It includes Calculator windows to help you work

Transistor Switches

Transistors are used in one of two ways:

- as switches
- as amplifiers.

Transistor amplifiers are described later in the book. Transistor switches are described in this topic, using three different types of transistor.

As we shall explain, the purpose of most transistor switches is to control an electrical device such as a lamp, a siren or a motor.

The switched device usually requires a current of several milliamperes, possibly several amperes. The current needed for operating the transistor switch is a lot smaller, often only a few microamperes. This makes it possible to control high-current devices from sensors, logic gates and other circuits with low-current output. The main limitation is that only devices working on direct current can be switched, but not devices powered by alternating current. These points are illustrated by the examples below.

SWITCHING A LAMP

The switching component in this circuit is a **metal oxide silicon field effect transistor**. This name is usually shortened to MOSFET. The words 'metal oxide silicon' refer to the fact that the transistor consists of a metal conductor, an insulating layer of oxide (actually silicon oxide), and a semiconducting layer of silicon.

The words 'field effect' refer to the way in which this type of transistor works. It works by placing an electric charge on the 'metal' part of the transistor, known as the **gate**. The charge on the gate results in an electric field, and the effect of this field is to control the amount of current flowing through the semiconductor layer. Current flows from the **drain** terminal to the **source** terminal.

FIGURE 2.1   A MOSFET has three terminals: gate, drain and source.

FIGURE 2.2   This medium-power MOSFET is rated to switch currents up to 13 A. Note the metal tag for attaching a heat sink.

Drain current (I_D) flows from the drain to the source, if:

- the drain terminal is positive of the source terminal and,
- the gate is charged to a voltage that is sufficiently positive of the source.

In this way, the MOSFET acts as a **voltage-controlled switch**. The voltage at which the transistor switches on is called the **threshold voltage**.

A MOSFET of the type just described is known as an **n-channel enhancement MOSFET**. There are other types (pp. 16 and 69), but this type is by far the most commonly used.

SENSOR

The light-sensitive component in this circuit is a **photodiode**. Like any other diode, it conducts in only one direction.

FIGURE 2.3   The can of this photodiode has a lens top. The diode is visible through this as a square chip of silicon.

Usually a photodiode is connected with **reverse bias**. This would mean that no forward current flows through it. However, a small **leakage current** passes through it. The leakage current is only a few nanoamps in darkness, but rises to several microamps when light falls on the photodiode.

FIGURE 2.4   Leakage current passes through the photodiode only in the light.

SWITCHING CIRCUIT

This circuit uses a MOSFET transistor for switching a lamp. We might use this to control a low-voltage porch lamp which is switched on automatically at dusk.

The sensor is a photodiode, which could be either of the visible light type or the type especially sensitive to infra-red radiation. It is connected so that it is reverse-biased. Only leakage current passes through it. The current varies according to the amount of light falling on the diode. In darkness, the current is only about 8 nA but it rises to 3 mA or more in average room lighting.

FIGURE 2.5   The MOSFET switches on the filament lamp when the light intensity falls below a given level.

The current change is converted into a voltage change by resistor R1. In light conditions, the current is relatively large, so the voltage drop across the resistor is several volts. The voltage at the point between R1 and the diode is low, below the threshold of the transistor. The transistor is off and the lamp is not lit.

At dusk, light level falls, the leakage current falls, and the voltage drop across R1 falls too. This makes the voltage at the gate of the transistor rise well above its threshold, which is in the region of 2.4 V. The transistor turns fully on. We say that it is **saturated**. When the transistor is 'on', its effective resistance is only 5 Ω. Current flows through the lamp and transistor, lighting the lamp.

The current flowing down through the sensor network (R1 and D1) is exceedingly small — no more than a few microamps. Little current is available to charge the gate of Q1. However, a MOSFET has very high input resistance, which means that it requires virtually no current to turn it on. It is the *voltage* at the gate, not the current, that is important.

FIGURE 2.6 A breadboarded version of the light-sensitive MOSFET switch. The photodiode and resistor are on the left. The low-power MOSFET is the small black package on the right.

Because the gate requires so little current, a MOSFET is the ideal type of transistor for use in this circuit. If we were to use a BJT, we might find that the sensor network was unable to provide enough current to turn it on.

The lamp specified for this circuit has a current rating of 100 mA. The transistor has a rating of 270 mA, so it is not in danger of burning out. There is no need to use a power MOSFET.

VARIATIONS

A useful improvement is to reduce R1 (to 820 kΩ) and wire a variable resistor (about 1 MΩ) in series with it. Connect the gate of Q1 to the wiper of the variable resistor. This allows the switching level of the circuit to be adjusted.

The circuit may be made to operate in the reverse way by exchanging the resistor and photodiode. It may be necessary to alter the value of the resistor.

It can be seen that block diagrams of this kind represent the *flow of information* through a system.

A circuit such as this does not normally need a heat sink on the transistor, even if the current being switched is several amps. The circuit is always in either of two states, 'off' or 'on'. When the transistor is off, there is no current and no heating effect. When the transistor is on, it is *fully* on. Its resistance may be only an ohm or two, or even less. Little energy is dissipated within the transistor and it never gets hot.

SYSTEM DESIGN

Electronic systems often consist of three stages. The first stage is the **input** stage. This is the stage at which some feature of the outside world (such as a temperature, the light level, or a sound) is detected by the system and converted into some kind of electrical signal.

The last stage is the **output** stage in which an electrical signal controls the action of a device such as a lamp, a motor, or a loudspeaker.

Connecting the input and output stages is the **signal processing** stage. This may comprise several sub-stages for amplifying, level-detecting, data processing, or filtering.

As a system, this circuit is one of the simplest. It has three sections:

FIGURE 2.7

The **input** section consists of the photodiode.

The **processing** section consists of R1 and Q1. R1 generates a voltage, which is fed to the gate of Q1, which is either turned 'off' or 'on'.

The **output** section consists of the lamp, which is switched by Q1.

Three-stage systems are very common. There are more examples in this topic and elsewhere in this book.

LIGHT-SENSITIVE ALARM

Compared with the circuit on p. 12, this circuit uses a different type of light sensor, and a different type of transistor. It has a similar three-stage system diagram:

FIGURE 2.8

- **Input** is accepted by a **light-dependent resistor** (or LDR). This is made from a semiconductor material such as cadmium sulphide. The resistance of this substance varies according the amount of light falling on it. The resistance of an LDR ranges between 100 Ω in bright light to about 1 MΩ in darkness.
- **Processing** uses a voltage divider (R1/R2) that sends a varying voltage to the **bipolar junction transistor**, Q1. The voltage at the point between the resistors is low (about 10 mV) in very bright light, but rises to about 5 V in darkness. With LDRs other than the ORP12, voltages will be different and it may be necessary to change the value of R1 or replace it with a variable resistor.
- **Output** is the sound from an **audible warning device**, such as a siren.

FIGURE 2.9 The audible warning device sounds when the light intensity falls below a given level.

The function of the circuit is to detect an intruder passing between the sensor and a local source of light, such as a street light. When the intruder's shadow falls on the sensor, its resistance increases, raising the voltage at the junction of the resistors. Increased current flows through R3 to the base of Q1. This turns Q1 on and the siren sounds.

FIGURE 2.10 The quadruple siren to the right of the breadboard emits an unpleasantly loud scream when the LDR on the left is shaded.

This circuit uses a BJT; almost any npn type will do, provided it is able to carry the current required by the audible warning device, which in this example is a piezo-electric siren. A typical ultra-loud (105 dB) warbling siren suitable for use in a security system takes about 160 mA when run on 12 V. The BC337 transistor is rated at 500 mA.

BJT SWITCHES

Like a MOSFET, a BJT has three terminals. Because the BJT works differently, they have different names: base, collector and emitter.

A BJT is connected so that its collector terminal is several volts positive of the emitter terminal. When current (I_B) flows into the base terminal, a much bigger current (I_C) flows into the collector terminal. The combined currents I_B and I_C flow out of the emitter terminal.

The combined currents make up the emitter current I_E, and:

$$I_E = I_B + I_C$$

The action of a BJT depends on the fact that I_C is alway very much larger than I_B. For a typical BJT it is about 100 times larger. Thus, small variations in a small current result in large variations in a large current.

When used as a switch, a BJT is a **current-controlled switch**. This contrasts with a MOSFET, which makes a voltage-controlled switch. The amount of current needed to switch on the transistor in this circuit depends on the amount of **collector current** needed to drive the siren. This depends on the **current gain** of the transistor.

The siren used in the demonstration circuit requires 160 mA. The BC377 has a current gain of between 100 and 600. Taking the minimum gain of 100, the base current needs to be:

$$I_B = I_C/\text{gain} = 160/100 = 1.6 \text{ mA}$$

Inside an npn BJT there is a pn junction between the base and emitter terminals. This junction has the properties of a diode. In particular, when it is forward biased, there is a voltage drop of about 0.7 V between the base and emitter terminals. In practice, the voltage drop may be 0.5 V to 0.9 V, depending on the size of I_B. Therefore, when Q1 is in the 'on' state, its base is, on average, 0.7 V positive of its emitter.

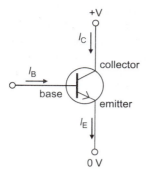

FIGURE 2.11 In a BJT, a small current controls a much larger current.

Now we can calculate what voltage from the R1/R2 potential divider is needed to switch on Q1 and sound the siren. We need a minimum of 1.6 mA flowing through R3. Given that R3 is 82 Ω, the voltage across R3 must be:

$$\text{current} \times \text{resistance} = 1.6 \times 10^{-3} \times 82 = 1.31 \text{ V}$$

If the voltage at the R1/R2 connection is less than 1.31 V, no current flows through R3 and Q1 is off. I_C is zero and the siren does not sound. If the voltage is greater than 1.31 V, Q1 turns on and becomes **saturated**. I_C rises to 160 mA and the siren sounds. There is a state between Q1 being off and being on and saturated, but we will leave discussion of this until later.

This circuit is able to use a BJT because the sensor network is able to deliver a current of the size required to turn on Q1. It is also possible to use a MOSFET. In this case, R3 is not required as almost no current flows to the gate of a MOSFET.

HEATER SWITCH

The heater switch is a circuit for maintaining a steady temperature in a room, a greenhouse, or an incubator. The sensor is a **thermistor**, a semiconductor device. Its resistance decreases with increasing temperature, which is why it is described as a **negative temperature coefficient** (ntc) device. Thermistors are not ideal for temperature *measuring* circuits, because the relationship between temperature and resistance is far from linear. This is no disadvantage if only one temperature is to be set, as in this example.

Thermistors have the advantage that they can be made very small, so they can be used to measure temperatures in small, inaccessible places. They also quickly come to the same temperature as their surroundings or to that of an object with which they are in contact. This makes them good for circuits in which rapid reading of temperature is important.

In many applications, these advantages offset the disadvantage of non-linearity.

The switched device in this example is a **relay**. This consists of a coil wound on an iron core. When current passes through the coil the core becomes magnetised and attracts a pivoted armature. The armature is pulled into contact with the end of the core and, in doing so, presses two spring contacts together. Closing the contacts completes a second circuit which switches on the heater. When the current through the coil is switched off, the core is no longer magnetised, the armature springs back to its original position, and the contacts separate, turning off the heater.

FIGURE 2.12 A relay is necessary to switch on a heater which runs from the 230 V AC mains.

Most relays have contacts capable of switching high voltages and currents, so they may be used for switching mains-powered devices running on alternating current. Current ratings of various types of relay range from 1 A up to 40 A. Various arrangements of switch contacts are available, including simple **change-over contacts** (see drawing above) and relays with three or more independent sets of contacts. Changeover contacts are normally open (no) or normally closed (nc). They can be wired so that a device can be turned off at the same time that another is turned on. Relays are made in an assortment of sizes, including some only a little larger than a transistor, suitable for mounting on printed circuit boards.

Self Test

What is the minimum contact rating for a relay required to switch a 1000 W, 230 V floodlamp?

WARNING

The circuit above right illustrates the switching of a device powered from the mains. **This circuit should not be built or tested except by persons who have previous experience of building and handling mains-powered circuits.**

THERMISTOR

Thermistors are available in several resistance ratings, the one chosen for this circuit having a resistance of 1 kΩ at 25°C. VR1 is adjusted so that the voltage at point A is close to 0.6 V when the temperature is close to that at which the heater is to be turned on. The transistor is just on the point of switching on. As temperature falls further, the resistance of the thermistor R1 increases (see p. 356), raising the voltage at A, and increasing i_B, the base current. This switches on the transistor and current passes through the relay coil. The relay contacts close, switching on the heater.

Note that the heater circuit is electrically *separate* from the control circuit. This is why a relay is often used for switching mains-voltage alternating current.

After a while, as the heater warms the thermistor, R1 decreases, the voltage at point A falls and the transistor is turned off. The relay coil is de-energised and the heater is turned off.

This kind of action is called **negative feedback**. It keeps the temperature of the room more or less constant. The action of the circuit, including the extent to which the temperature rises and falls, depends on the relative positions of the heater and thermistor and the distance between them.

PROTECTIVE DIODE

The diode D1 is an important component in this circuit.

At (a) below a transistor is controlling the current through an inductor. When the transistor switches off (b), current ceases to flow through the coil and the magnetic field in the coil collapses. This causes an **emf** to be induced in the coil, acting to *oppose* the collapse of the field. The size of the emf depends on

how rapidly the field collapses. A rapid switch-off, as at (b), collapses the field instantly, resulting in many hundreds of volts being developed across the transistor. This emf produces a large current, which may permanently damage or destroy the transistor.

Emf

This stands for electromotive force. Emf is a potential difference produced by conversion of another form of energy into electrical energy. Ways of producing emf include chemical (dry cell), mechanical (dynamo, rubbing a plastic rod with a woollen cloth), and magnetic (collapsing field, car ignition system).

If a diode is wired in parallel with the inductor, as at (c), it protects the transistor by conducting the induced current away to the power line. It is a **protective diode**.

INDICATOR LAMP

An indicator lamp is an optional but useful addition to this heater circuit. It could also be added to the other switching circuits we have studied. The idea is to have a lamp that comes on when the circuit is switched on. We could use a filament lamp (such as a 6 V torch bulb) but a light-emitting diode is generally preferred. This is because, for a given brightness, LEDs do not take as much current as filament lamps. Also, filament lamps do not last as long as LEDs. An LED needs a current of a few tens of milliamps. Often 20 mA or 25 mA makes the LED shine brightly. Like any other diode, an LED has a forward voltage drop across it when it is conducting. For most LEDs the drop is about 2 V.

FIGURE 2.13

FIGURE 2.14 An LED normally has a resistor in series with it.

The LED will burn out if the voltage across it is much more than 2 V. If the LED is being powered by a supply greater than 2 V, we need a resistor (R3) in series with it. This drops the excess voltage.

In the diagram above, the supply is 6 V, so the excess voltage is 4 V. If we decide to have a current of 25 mA through the LED, the same current passing through R3 must produce a drop of 4 V across it. So its resistance must be:

$$R_3 = 4/0.025 = 160 \ \Omega$$

Select the nearest standard value. In this case, a 160 Ω resistor can be used, though a 150 Ω resistor would be close enough.

Self Test

1 Calculate the value of the series resistor when:
  a the supply voltage is 9 V and the current required is 20 mA.
  b the supply voltage is 24 V and the current required is 15 mA.
2 If the supply is 12 V and the series resistor is 1.5 kΩ, what current flows through the LED?

In general, if the supply voltage is V_S and the current is to be i_{LED}, the series resistor R_S is:

$$R_S = (V_S - 2)/i_{LED}$$

Certain types of LED, such as flashing LEDs, have built-in resistance and do not need a series resistor.

OVERHEATING ALERT

Previous circuits in this topic have used BJTs of the type known as **npn** transistors. Here we use the other type of BJT, a **pnp transistor**. The action of a pnp transistor is similar to that of an npn transistor. In particular, a small base current controls a much larger collector current. However, the polarities are reversed:

- The transistor is connected with its emitter *positive* of its collector.
- Base current flows *out of* the base terminal instead of flowing into it.
- When the transistor is 'on', the base terminal is 0.7 V *negative* of the emitter.

Here is a switching circuit using a pnp BJT:

FIGURE 2.15 Compare this circuit with that in p. 14. Note the symbol for a pnp transistor.

The circuit is set by adjusting VR1 until the LED is just extinguished. The base (b) potential is then a little above 11.3 V (that is, the base-emitter voltage drop is a little less than 0.7 V). If the thermistor (R2) is warmed, its resistance decreases. This reduces the voltage at the point between R1 and R2. Current flows out of the base, turning the transistor on. When the transistor is on, the collector current flows in through the emitter (e), out of the collector (c) and on through the LED and series resistor R3. Thus, the LED comes on whenever the temperature of R1 rises. By adjusting VR1, the circuit can be set to switch on the LED at any required temperature in the region of 25°C.

INVERSE ACTION

The switching action is the opposite to that of the npn transistor circuit on p. 14. There the relay comes on when temperature of the thermistor falls. In this circuit the LED comes on as temperature rises.

We could have achieved the equivalent action with an npn transistor in the relay circuit by exchanging VR1 and the thermistor. The values of the resistors would have needed altering but the action would be to turn on the relay with *falling* temperature.

PNP OR NPN?

The inverse action can be achieved either by exchanging components or by using a pnp transistor.

This illustrates a point about electronic circuits, that there are often two or more ways of doing the same thing. Sometimes both ways are equally effective and convenient. In other instances we may have reason to prefer one way to the other.

In practice, npn transistors are used far more often than pnp transistors. Although sometimes a resistor or logic gate may be saved in a circuit by using pnp, it is usually just as convenient to swap components to invert the action.

The main reason for using pnp is when we want a given signal to produce an action and its inverse *at the same time*. Then we use a pair of transistors, npn and pnp, with equal but opposite characteristics. Examples are seen in the power amplifiers described on pp. 124−125.

MOSFET TYPES

MOSFETs too are made that work with polarities opposite to those of the n-channel MOSFETs described in this topic. These are known as **p-channel MOSFETs** (p. 69).

SCHMITT TRIGGER

In the two previous circuits, the resistance of the thermistor changes relatively slowly, so that the circuit spends several minutes in a state intermediate between 'off' and 'fully on'. The transistor is partly, not fully, switched on. In the heater switch circuit,

the coil of the relay is not fully energised and the armature tends to vibrate rapidly, switching the heater on and off several times per second. This leads to sparking at the contacts, which may become fused together so that the relay is permanently 'on'. In an LED switching circuit, the LED gradually increases in brightness as temperature increases. It gives no clear indication of whether or not the temperature is too high. The circuit below avoids these problems, giving a sharp trigger switching action.

This **Schmitt trigger** consists of two npn transistors. Q1 is switched by the sensor network. Q2 is switched by Q1 and in turn switches the relay.

The thermistor and variable resistor are arranged so that the voltage at point A *falls* as temperature falls. This is the opposite action to that in the circuit on p. 15.

This is necessary because we have *two* inverting transistors in this circuit. The resistance of R3 must be greater than that of the relay coil, so that more current flows through R6 when Q2 is on, and less flows when Q1 is on.

SNAP ACTION

As temperature falls and Q1 begins to turn off, Q2 begins to turn on. The increasing current through R6 raises the voltage at the Q1 emitter. As the Q1 base voltage continues to fall its emitter voltage starts to rise, rapidly reducing the base current and turning Q1 off very rapidly. If Q1 turns off faster, then Q2 turns on faster. The circuit spends much less time in its intermediate state.

FIGURE 2.16 The Schmitt trigger circuit has a 'snap' action, which is suitable for switching heaters, indicator lamps and many other devices.

HYSTERESIS

In this circuit the temperature at which the relay is turned on or off depends on whether the temperature is rising or falling.

The action of the circuit is such that the 'turn-on' temperature is lower than the 'turn-off' temperature. We call this action **hysteresis**.

The circuit works differently in the two states:

- **The heater is off**, the temperature of R1 is relatively high but falling, and its resistance is low, but rising. The voltage at A is relatively high but falling. Q1 is on and Q2 is off. Current through R6 and the voltage across it are fairly small. For example, measurements on this circuit found the voltage at B to be 0.4 V. This is due to the current flowing through R3 and Q1 and finally through R6. As the voltage at A falls to 1.1 V (0.4 V + 0.7 V due to the base-emitter voltage of Q1), the circuit starts to change state, turning the heater on.
- **The heater is on**, and this increases the current through R6 because the resistance of the relay coil is lower than that of R3. The increased current makes the voltage at B rise to, say, 0.9 V. Now the voltage at A has to rise to 1.6 V (0.9 V + 0.7 V due to v_{BE}) before Q1 can be turned on again and the circuit can return to its original state.

In terms of heating, the heater comes on when the room has cooled to a set temperature, say 15°C. We call this the **lower threshold**. The heater stays on, perhaps for 10−15 minutes, until the room has warmed to a higher temperature, say 17°C, then goes off. We call this the **higher threshold**. The difference betweeen the two thresholds, in this case two degrees, is the hysteresis of the circuit. This is a much better action than having the heater continually switching on and off several times a minute at each little rise or fall in temperature.

Any of the switching circuits in this topic could be modified to have a Schmitt trigger between the sensor and the switched device.

FIGURE 2.17 A Schmitt trigger is one of the processing stages of this light-sensitive alarm.

LOGICAL CONTROL

The input of the circuit below is taken from the output of a logic gate (p. 149). This would not be able to supply enough current to light the lamp, but can raise the voltage at the gate of the MOSFET sufficiently to turn the transistor on. When the logic output is low (0 V) the voltage at the gate of Q1 is below the threshold (2.5 V) and the transistor is fully off. There is no drain current and the lamp is unlit.

FIGURE 2.18 This is how a lamp may be controlled by a logic circuit, such as a microcontroller.

When the logic output goes high (say, +5V for a TTL output) this puts the gate of Q1 well above the threshold. The transistor is turned fully on and the maximum drain current flows, turning the lamp on.

The power supply for the lamp may be at a higher voltage than that used to supply the logic circuit. If the logic is operating on 5V, the lamp or other device could be powered from, say, a 24V DC supply, provided that the current rating (0.5 A) and power rating (1 W) of the transistor are not exceeded.

This circuit could also be used to switch other devices, including a **solenoid**. A solenoid is a coil wound on a former. There is an iron **plunger** or **armature** which slides easily in and out of the coil. Usually, the plunger is held by a light spring so it rests only partly inside the coil when there is no current flowing (see photo, on the next page).

When current flows through the coil, the magnetic field strongly attracts the plunger, which is drawn forcefully into the coil.

There is usually some kind of mechanism linked to the plunger. For example, the plunger may be attached to a sliding bolt, so that energising the coil causes a door to be bolted. In this way the door may be bolted by computer control.

As with the relay circuit on p. 15, a solenoid switching circuit requires a diode to protect the switching transistor.

FIGURE 2.19 When a current passes through the coil of this solenoid, the plunger moves strongly in the direction shown by the arrow.

EXTENSION BOX 4 Current Gain

The gain of a BJT is expressed as the ratio of the collector current to the base current. This is known as the **large signal current gain**, h_{FE}:

$$h_{FE} = I_C/I_B$$

Because h_{FE} is a current divided by a current, it has no unit. It is just a *number*.

Example

When the base current to a given BJT is 270 μA, the collector current is 40.5 mA. What is the current gain of the BJT?

$$h_{FE} = I_C/I_B = 40.5/270 \ [\text{mA}/\mu\text{A}] = 150$$

The current gain is 150.

Self Test

If the gain of a BJT is 240, and I_B is 56 μA, what is I_C?

Memo

There is also the **small signal current gain**, h_{fe} (small letters for the suffix 'fe'). A *small change* i_b in the base current causes a relatively *small change* i_c in the collector current. Then, $h_{fe} = i_c/i_b$.

In most cases, h_{fe} has a value equal or close to that of h_{FE} (see above).

EXTENSION BOX 5 Transconductance

A MOSFET is a voltage controlled device. Its 'gain' is expressed as the *change* in the drain current for a given *change* in gate-source voltage. This is known as the **transconductance**, g_m:

$$g_m = i_d/v_{gs}$$

Because g_m is a current divided by a voltage it has the units of conductance (the inverse of resistance). The unit of conductance is the **siemens**, symbol **S**.

Example

When the gate-source voltage of a given MOSFET increases by 2 V, the drain current increases by 1.7 A. What is the transconductance of the MOSFET?

$$g_m = i_d/v_{gs} = 1.7/2 = 0.85 \ \text{S}$$

The transconductance is 0.85 S, or 85 mS.

Note that the calculations in Box 4 and this box assume that the collector-emitter pd and the drain-source pd are constant, and are sufficient to drive the current through the transistor.

DATA — Bipolar Junction Transistors

The transistors are listed by **type number**. Many of the circuit diagrams in this book specify type number BC548. This is an easily available, cheap, general-purpose BJT, made by several different manufacturers. As can be seen from the table, there are several other types of BJT that would work in these circuits just as well as the BC548.

The type number is not particularly useful when you are selecting a transistor. The easiest way is to refer to the final column of the table and select by application. The majority

of transistors are described as 'general-purpose' (GP) possibly with some additional features mentioned. One of these types will suit most projects.

All transistors in this table are npn except the BC557, which is pnp.

Probably the most important item is the **maximum collector current**. The data sheet lists this in a column headed **I_C (max)**. This is the maximum collector current (almost equal to the emitter current) that the BJT can pass when it is saturated. It is equal to the current passing through the load. The next column lists the **maximum power rating,** P_{tot}.

The **current gain** specified in the table is the small signal current gain, h_{fe}, mentioned on p. 20. The value of h_{FE} is almost the same as h_{fe}, so these values can be used in transistor switch calculations.

It is measured with the collector current at the stated value. Usually, a range is specified because transistors of the same type may vary widely in gain. The BC548, for instance, has a gain of 110 to 800. Sometimes transistors are graded by the manufacturer and sold in gain groups. A transistor sold as a BC548A, for instance, has a gain in the range 110 to 220. BC545B ranges from 200 to 450, and BC548C ranges from 420 to 800.

The fourth important feature of a transistor is its **maximum frequency,** F_t. A transistor being used as a switch responds quickly enough for most applications.

When working at radio frequencies, it may not be able to respond quickly enough to changes in the signal current. The gain falls off at high frequencies. F_t is defined as the frequency for which the gain falls to 1. For many transistors, F_t is rated in hundreds of megahertz and, in the table, is quoted in megahertz at a given collector current.

Normally, we do not have to worry about F_t. It is mainly in radio circuits and computer circuits that such high frequencies are used. But it can be seen from the table that high-power transistors have much lower F_t than other types.

Type	Case	I_c (max)	P_{tot}	h_{fe} at I_c	(mA)	F_t at I_c	(MHz at mA)	Applications
BC107	TO-18	100 mA	300 mW	110−450	2	300	10	GP small signal amplifier
BC108	TO-18	100 mA	300 mW	110−800	2	300	10	GP small signal amplifier
BC109	TO-18	100 mA	300 mW	200−800	2	300	10	Low noise, small sig. amp.
BC109C	TO-18	100 mA	300 mW	420−800	2	300	10	Low noise, high gain, amp
BC337	TO-92	500 mA	625 mW	100−600	2	100	10	Output stages
BC548	TO-92A	100 mA	500 mW	110−800	2	300	10	Low noise, small sig. amp.
BC557	TO-92	100 mA	625 mW	110−800	2	280	10	pnp output transistor
BC639	TO-92	1 A	1 W	40−250	150	200	10	Audio output
BD139	TO-126	1.5 A	8 W	40−250	150	250	50	Power GP
BD263	TO-126	4 A	36 W	750	1.5 A	7	1.5 A	Med. power Darlington
BUX80	TO-3	10 A	100 W	30	1.2 A	8		High current switch
MJE3055	TO-220	10 A	75 W	20−70	4 A	2	500	Power output
2N2222	TO-18	800 mA	500 mW	100−300	150	300	20	High speed switch
2N3641	TO-105	500 mA	350 mW	40−120		250	50	GP, amplifier, switch
2N3771	TO-3	30 A	150 W	15−60	1.5 A	0.2	1 A	Power output
2N3866	TO-39	400 mA	1 W	10−200	50	500	50	VHF amplifier
2N3904	TO-92	200 mA	310 mW	100−300	10	300	10	Low power GP

DATA − MOSFETs

The table below lists a selection of low-cost, low-voltage MOSFETs. Maximum drain-source voltages range between 60 V and 100 V, except for the IRF630 which is 200 V.

The most useful figures for choosing a transistor are maximum drain current and drain-source source resistance when 'on'.

The threshold voltage of a given type may vary considerably between different transistors of the same type. Usually it is between 2 V and 4 V. This is why we have not quoted threshold voltages in this table.

Transconductance is of less important when selecting, but you will need it when calculating outputs.

In general, transistors with low I_D and low g_m are considerably cheaper. In this table, the cheapest n-channel MOSFET is the ZVN3306A and the dearest is the IRFD024.

Type	Case	n/p	I_D (max)	P_{tot}	R_{DS} (max)	g_m (S)
BUZ73L	TO-220	n	7 A	40 W	0.40 Ω	5.00
IRF630	TO-220	n	9 A	75 W	0.40 Ω	3.00
IRF9540	TO-220	p	19 A	150 W	0.20 Ω	5.00
IRFD110	TO-250	n	1 A	1.3 W	0.54 Ω	1.30
IRFD024	TO-250	n	2.5 A	1.3 W	0.1	0.90
ZVN2106A	Eline	n	450 mA	1.5 W	2 Ω	0.40
ZVN3306A	Eline	n	270 mA	625 mW	5 Ω	0.15
ZVP2106A	Eline	p	280 mA	700 mW	5 Ω	0.15
2N7000	TO-92	n	200 mA	400 mW	5 Ω	0.10

> **BJTs or MOSFETs?**
>
> When choosing between BJTs and MOSFETs, keep these points in mind:
>
> BJTs need appreciable current to drive them; MOSFETs need none.
>
> For switching: MOSFETs have very low 'on' resistance and very high 'off' resistance.
>
> For switching: MOSFETs are faster (but BJTs are usually fast enough).
>
> As amplifiers: BJTs have linear input/output relationship. MOSFETs are not perfectly linear, which may cause distortion of large signals.
>
> MOSFETs need no resistor in their gate circuit, which simplifies construction.
>
> MOSFETs are liable to damage by static charge. They need careful handling.

ACTIVITIES — SWITCHING CIRCUITS

1 On a breadboard or on stripboard, build the switching circuits illustrated in this topic and test their action. Do not connect any of the relay circuits to the mains. Instead, use a power pack running at, say, 15 V, to power the switched device.

2 There are a number of simple ways in which these circuits can be altered to change their action:
   - Modify the lamp switching circuit (p. 10) so that the lamp is turned on when light level increases.
   - Add a variable resistor to the circuit on p. 14 to make it possible to set the light level at which the switch operates.
   - Try increasing and decreasing the resistance of R6 in the circuit on p. 18 to see what effect this has on the hysteresis.
   - Look at the effect of changing the value of R3.
   - Build a flip-flop based on a pair of CMOS NAND gates and use this to drive a MOSFET and filament lamp.

3 Use the circuits as a guide when designing and building switching circuits with the following functions:
   - A photodiode and a BJT to switch a lamp on when light falls below a set level.
   - A light dependent resistor and a MOSFET to switch an LED on by detecting the headlamps of an approaching car at night.
   - A circuit to sound a siren as a garden frost warning.
   - A circuit to switch on an electric fan (a small battery-driven model) when the temperature falls below a given level. This circuit is better if it has hysteresis.

   - Any other switching function that you think is useful.

4 When you have designed a circuit, either build it on a breadboard, or 'build' and test it on a circuit simulator, using a computer.

QUESTIONS ON TRANSISTOR SWITCHES

1 Describe the action of a MOSFET switch used with a photodiode as sensor for switching on a filament lamp.

2 Design a fire-alarm circuit for switching a warning lamp on when the temperature rises above a given level.

3 Design a "Dad's home!" circuit for switching on a small piezo-electric sounder when the sensor detects approaching car headlamps at night. Calculate suitable component values.

4 Explain why it is essential to include a protective diode in a circuit that switches an inductive device.

5 Describe the operation of a simple changeover relay, illustrating your answer with diagrams.

6 Draw a diagram of a Schmitt trigger circuit based on BJTs and explain how it works.

7 Design a Schmitt trigger circuit based on MOSFETs.

8 Design a circuit for operating a magnetic door release automatically when a person approaches the doorway. The release operates on a 12 V supply and requires 1.5 A to release it.

9 Compare the features of BJTs and MOSFETs when used as switches.

10 What is a light dependent resistor? What is the effect of shining light on it? How can an LDR be used to operate a BJT switch to turn on a lamp at night, when it is dark?

11 Why is a thermistor unsuitable for a use in a temperature measuring circuit?

12 On p. 21 are data tables for BJTs and MOSFETs. Use these tables to select transistor types suitable for the switching circuits listed on this page. Access manufacturers' catalogues on CD-ROM or the Internet if you need further information. For each circuit, give your reasons for selecting the transistor. You can assume that all the transistors listed are suitable for the operating voltages of the listed circuits:

   a Select alternative transistors for the switching circuits described in this topic. Note the modifications, if any, to component values that may be necessary to enable the circuit to work with the chosen transistor.

   b A light-dependent resistor is the sensor for a circuit in which a MOSFET switches a current of

20 mA through an LED. The power supply is 6 V. Select a suitable MOSFET and design the circuit.

c The circuit is a morning 'wake-up' alarm. A photodiode is the sensor and the circuit sounds an electronic buzzer when the light intensity rises above a given level. The buzzer takes 15 mA at 6 V. Select a BJT or MOSFET and design the circuit.

d A frost alarm uses a thermistor as sensor and a BJT to turn on a filament lamp when the temperature falls below a given level. The 3.6 W lamp runs on 12 V. The switching temperature is adjustable. Select the transistor and design the circuit.

e A 6 V motor takes 300 mA and is to be switched on by a high output level (5 V) from a logic circuit. Select the transistor and design the circuit.

f The water level in a tank is monitored by a pair of probe wires placed closely side by side. When the probes are covered the resistance between them is 1 MΩ. The input stage of the system is shown below. Find a suitable value for R1, and design the remainder of the circuit.

FIGURE 2.20 The input stage of the circuit for question 12f.

g A greenouse window is opened by a solenoid that runs on a 12 V supply. The coil resistance is 14 Ω. Design a switching circuit to energise the solenoid when the temperature rises above a given level.

As further work on these questions, build each circuit you have designed, to check that it works.

EXTENSION QUESTIONS

13 Why is the transconductance of a MOSFET expressed in siemens, but the gain of a BJT has no unit?

14 A MOSFET has a g_m of 500 mS. If the gate-source voltage increases by 50 mV, what is the change in I_D?

15 A BJT is to switch on a small DC motor that takes 400 mA at 12 V. The current gain of the transistor is 250. What minimum base current is needed?

16 Given a MOSFET with a g_m of 1.6 S, what change of gate voltage is needed to decrease I_D by 400 mA?

MULTIPLE CHOICE QUESTIONS

1 When used as a switch, a BJT is connected in:
   A the common-emitter connection.
   B the common-collector connection.
   C the common-base connection.
   D none of the above.

2 Which of these features of a MOSFET is not of importance in a transistor switch?
   A high input resistance.
   B low 'on' resistance.
   C medium transconductance.
   D low input capacitance.

3 A light-dependent resistor is usually made from:
   A silicon oxide.
   B p-type semiconductor.
   C cadmium sulphide.
   D carbon.

4 A thermistor is not ideal for a temperature measuring circuit because:
   A its resistance is too high
   B its response is not linear.
   C it responds too slowly to changes of temperature.
   D it has a negative temperature coefficient.

5 The maximum current that a MOSFET can switch when operating on a 9 V supply and with an 'on' resistance of 2 Ω :
   A depends on the device being switched.
   B depends on its maximum current rating.
   C is 4.5 A.
   D is 222 mA.

6 The voltage level at which a Schmitt trigger switches on is different from that at which it switches off. We call this:

   A negative feedback
   B current gain
   C hysteresis
   D positive feedback.

Transistor Switches

The Companion site (see p. ix) has several animated diagrams that illustrate the action of transistor switches.
   Visit the site and see how these circuits work.

8 When a capacitor is used to transfer a signal between two parts of a circuit that are operating at different voltages, the capacitor is said to be:

    A coupling the circuits.
    B filtering the signal.

    C decoupling the circuits.
    D amplifying the signal.

There are more calculations on capacitance and impedance on the Companion website.

Using Capacitors

Topic 4 demonstrates that capacitors introduce time into the action of a circuit. The delay circuit below uses a capacitor in this way.

DELAY

The circuit below has a potential divider R2/R3 which has an output voltage 4.8 V. The operational amplifier (see Topic 11) compares this voltage with the voltage across C1.

The circuit is triggered by pressing S1 to discharge the capacitor fully. This brings the non-inverting input (+) of the op amp down to 0 V. The output of the op amp swings down to 0 V and the LED goes out.

When S1 is released, current flows through R1 and charges C1. As soon as the voltage across it reaches 4.8 V, the output of the op amp swings almost to 6 V and the LED comes on. With the values of R1 and C1 shown in the diagram, the delay is 75 s. Extension Box 9, on p. 49 shows how this result is calculated.

FIGURE 5.1   The time for which the LED is off depends on the values of R1 and C1.

Diagram Labelling

From now on, resistor and capacitor values are written in a shortened form on circuit diagrams.

MONOSTABLE MULTIVIBRATOR

This kind of circuit, usually known by the shortened name of **monostable,** is a useful pulse generator or pulse extender.

The monostable circuit (overleaf) consists of two transistor switches. The output from each switch is the input to the other. A system diagram shows that the connection from Q2 to Q1 is direct (through R3), but the connection from Q1 to Q2 passes through a delay stage. The delay stage is provided by C1 and R2 — another example of charging a capacitor through a resistor.

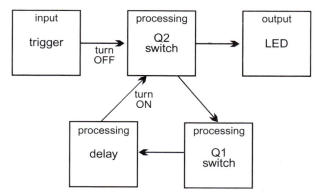

FIGURE 5.2

Without the delay, Q1 would turn Q2 on again immediately. You would not see even a flash of light from the LED. With the delay unit present, there is a delay of a few seconds while C1 charges through R2. Q2 is turned on when C1 has charged to the right level.

This circuit is stable when Q1 is off and Q2 is on. It remains indefinitely in that state. It is unstable in the reverse state, with Q1 on and Q2 off. After the delay, it goes back to the other state.

The behaviour of the circuit is more complicated than in the previous example, the base of Q2 being pulled down to a negative voltage when the circuit is triggered. The length of the pulse is approximately $0.7R_2C_1$. With the values in the diagram, the pulse length is $0.7 \times 6.8 \times 470$ [kΩ $\times$ μF] = 2.2 s.

Electronics: Circuits and Systems.

FIGURE 5.3 Cross-connected BJT switches form a monostable pulse generator.

The charging time t depends on the values of R1 and C1 according to the equation:

$$t = 1.1 R_1 C_1$$

$R_1 C_1$ is the time constant, and the factor 1.1 is obtained by a calculation similar to those set out on pp. 36–37. Note that t does not depend on supply voltage. This is one of the main assets of the 555, making it suitable for circuits powered by battery or other unregulated supplies.

555 TIMER IC

One of the most popular monostables is based on the 555 integrated circuit. It operates on a wide range of supply voltages, from 4.5 V to 16 V.

The original 555 uses bipolar circuitry, but there are several newer devices, notably the 7555, based on CMOS. The advantages of the CMOS versions is that they use far less current, produce longer timing periods, cause less disturbance on the power lines when their output changes state, and operate on a wider range of supply voltages (3 V to 18 V).

Bipolar devices: conduction is by electrons and holes, for example, in npn and pnp transistors.

The 7555 output is able to sink or source only 100 mA, in comparison with the 200 mA of the 555, but this is rarely a drawback. The 7555 has 2% precision compared with 1% for the 555.

Sink or source current: to take in or to provide current.

The 555 and 7555 are also available as dual versions, the 556 and 7556, with two timers contained in a 14-pin package.

The diagram on the right shows the 555 connected as a monostable. The length of the output pulse

depends on the time taken to charge the capacitor to two-thirds of the supply voltage.

The control pin is sometimes used to adjust the exact length of the period. If this facility is not required, the control pin is connected to the 0 V line through a low-value capacitor (for example 10 nF).

Self Test

1. A 555 monostable circuit has R1 = 2.2 kΩ and C1 = 1.5 μF. What is the length of the output pulse?
2. Design a 555 monostable circuit using a 220 nF capacitor to give an output pulse of 25 ms.

With the 7555, the control pin is left unconnected.

The 555 needs the decoupling capacitor C3 connected across the supply terminals, but decoupling is not necessary with the 7555.

FIGURE 5.4 The basic 555 monostable circuit. C2 and C3 are not required with the 7555 CMOS IC.

The trigger (set) and reset pins are normally held high. A low-going pulse (which needs only fall to two-thirds of the supply voltage) triggers the monostable. The device is not retriggerable, but the output pulse can be ended at any time by a low-going pulse on the reset pin. In the figure, the reset pin is permanently wired to the positive supply (high) so that resetting is not possible.

The action of the monostable is as follows:

1) **In the quiescent state**, the timing capacitor (C1) is charged to one-third of the supply voltage. The output is low (0 V). Current flowing through R1 after C1 is charged is passed to ground by way of the discharge pin. The threshold pin monitors the voltage across C1.

2) **When the circuit is triggered**, its output goes high and the discharge pin no longer sinks current. Current flows to the capacitor, which begins to charge.

3) **When the capacitor is charged** to two-thirds of the supply voltage, as monitored by the threshold pin, the output goes low. The charge on the capacitor is rapidly conducted away through the discharge pin until the voltage across the capacitor has fallen again to one-third of the supply voltage.

The 555 timer has the advantage that the voltage levels are precisely monitored by the internal circuitry, so that timing has high precision. Most often it is limited by the precision of the timing resistor and capacitor. This is particularly the case when an electrolytic capacitor is used to give a long period. They have wide tolerance (20%) and their capacitance varies with age and usage.

ASTABLE MULTIVIBRATOR

The 555 can be used as an astable simply by connecting the trigger pin to the threshold pin and adding another resistor. The circuit is triggered again as soon as the capacitor is discharged at the end of its cycle. The capacitor is charged and discharged indefinitely and the output alternates between high and low.

FIGURE 5.5 The 555 timer IC can also be used as an astable multivibrator.

The action of the 555 astable is divided into two phases.

1 **Charging:** Output is high while the capacitor is charged from one-third to two-thirds the supply voltage, through R1 and R2 in series. The time taken is:

$$t_H = 0.69(R_1 + R_2)C_1$$

$(R_1+R_2)C_1$ is the time constant for charging. The factor 0.69 is obtained by a calculation similar to those set out on pp. 36–37. Note that t does not depend on supply voltage.

2 **Discharging:** Output is low while the capacitor is discharged through R2, current flowing into the discharge pin. The time taken is:

$$t_L = 0.69\, R_2 C_1$$

The total time for the two phases is:

$$t = t_H + t_L = 0.69\, C_1(R_1 + 2R_2)$$

The frequency of the signal is:

$$f = \frac{1}{t} = \frac{1}{0.69\, C_1(R_1+2R_2)}$$

$$f = \frac{1.44}{C_1(R_1 + 2R_2)}$$

The **mark-space ratio** is the ratio between the period when output is high and the period when it is low. Using the equations for t_H and t_L we obtain the result:

$$mark - space\ ratio = \frac{t_H}{t_L} = \frac{R_1 + R_2}{R_2}$$

> **Self Test**
>
> A 555 astable has $R_1=10$ kΩ, $R_2=6.8$ kΩ and $C_1=150$ nF. What is the frequency and the mark-space ratio of the output signal?

It follows from this result that the mark-space ratio is greater than 1 in all circuits connected as in Figure 5.5. It is possible to obtain ratios equal to 1 or less than 1 by using diodes in the timing network to route the charging and discharging currents through different resistors.

ALTERNATING SIGNALS

So far in this topic, we have seen what happens when a capacitor is charged from a *constant* voltage it does not become completely discharged until the end of that half-cycle. As a result of these delays, the *peaks and troughs* of V_{OUT} coincide with the *ends* of the half-cycles of V_{IN}. Putting it another way, V_{OUT} lags about a quarter of a cycle behind V_{IN}.

CYCLES AND PHASE

It is conventional to measure differences of phase in terms of angle. When the sine of an angle is plotted, the curve is repeated every 360°. One complete cycle is equivalent to 360°, as illustrated in the diagram. We measure phase differences in terms of degrees of angle.

Angles are also measured in radians, where one complete cycle is 2π rad. So phase differences can also be expressed in radians. However, in this discussion we will use only degrees.

In the diagram, where the frequency is 1 kHz, the period of the waveform is 1 ms. So 1 ms is equivalent to 360°. On the graph, V_{OUT} can be seen to lag 0.2 ms behind V_{IN}. In terms of angle, this is equivalent to $0.2 \times 360 = 72°$. Yet we have said earlier that the lag is expected to be a quarter of a cycle, or 90°. The reason for the discrepancy is that the frequency is only *just* higher than the cut-off point, and that the flow of current does not exactly conform to the description.

The way that the phase lag varies with frequency can be shown by measuring the frequency response of the filter and plotting the phase lag over the range of frequencies:

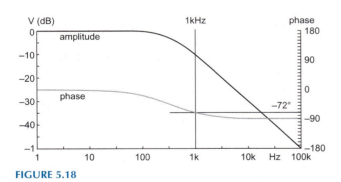

FIGURE 5.18

The plot of amplitude is typical for a lowpass filter. The −3dB point is 400 Hz. At 1 kHz, amplitude has begun to fall off.

The phase plot shows that there is little lag at low frequencies, where the period of the sinusoid is long compared with the time taken to charge and discharge the capacitor. Then the lag increases (the curve goes more negative) until it becomes −72° at 1 kHz. Above that, the lag increases and eventually settles to

−90° at frequencies higher than about 10 kHz. This agrees with the prediction made above.

CURRENT

From the discussion above, it is expected that the current flowing into and out of the capacitor will vary in a similar way to the output voltage. Here is a plot of V_{IN} and I_{OUT} against time:

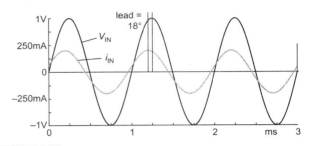

FIGURE 5.19

The plot shows that the curve of i_{OUT} is a sinusoid with frequency 1 kHz and amplitude 20 mA. The peaks and troughs of this curve are ahead of those of V_{IN}, with a phase difference of 18°. We say that current **leads** V_{IN} by 18°.

Current increases and begins to charge the capacitor as V_{IN} swings positive. However, the current curve begins to climb less rapidly as charge accumulates on the capacitor (compare with charging a capacitor, p. 36).

On the negative swing of V_{IN}, current reverses, to charge the capacitor in the opposite direction. Again, because charge accumulates on the capacitor, current reaches its peak earlier than V_{IN}.

Noting that $72° + 18° = 90°$, we would expect to find that current leads V_{OUT} by 90°. That this is so is shown in this plot:

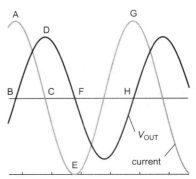

FIGURE 5.20   Current leads V_{OUT} by 90° in a lowpass RC filter.

Current is at its positive maximum (A) when it begins to charge the capacitor (B). Current falls as the capacitor gradually accumulates charge. Eventually, current is zero (C) when the capacitor is fully charged (D). Then the current reverses and the capacitor begins to discharge. Current reaches its negative maximum (E) as the discharged capacitor begins to charge in the opposite direction (F). The next half cycle is a repeat of the above, except that polarities are reversed. The current reaches its positive maximum again (G) as the capacitor begins to charge in the original direction.

HIGHPASS PHASE RESPONSE

Interchanging the resistor and capacitor reverses some of the effects described above. At low frequencies, V_{OUT} has a phase **lead** of 90°. This drops as frequency increases. At 1 kHz, with the values for R and C that we used before, the lead is 18°. It falls to zero at high frequencies.

V_{OUT} is the result of current flowing through the resistor to ground. By Ohm's Law, current is proportional to voltage at every instant. This means that the changes in current are exactly in phase with V_{OUT}.

ACTIVITIES — FILTERS

Investigate the action of some highpass filters set up on a breadboard or on a computer simulation. For example, use a 10 nF capacitor and a 3.9 kΩ resistor. Then repeat with other capacitors, such as 1 nF and 10 μF, and other resistors.

For a breadboarded filter, use a signal generator to supply a sinusoidal input signal at frequencies ranging from, say, 10 Hz to 1 MHz. Make the amplitude of the signal constant at, say, 1 V.

FIGURE 5.21　A breadboarded highpass RC filter, ready for testing.

As you vary the frequency of the input, use an oscilloscope to observe the output signal. Measure its amplitude at a series of frequencies, such as 10 Hz,

100 Hz, 1 kHz, and so on. In what way does amplitude vary with frequency?

Build or simulate low-pass filters and investigate their action in the same way as above.

QUESTIONS ON USING CAPACITORS

1　Design a circuit to switch an LED off for 0.5 s when a button is pressed (use an op amp or the 555 timer IC).

2　Design a circuit to switch an LED on for 20 s when a button is pressed (use an op amp or the 555 timer IC).

3　What is the action of a monostable multivibrator? Describe the circuit of a monostable.

Extension Box 9　Delays

Referring to the circuit on p. 41 and the calculations on p. 37, $V_S = 6$ and $v_C = 4.8$, which gives:

$$\frac{v_C}{V_S} = (1 - e^{-t/RC}) = \frac{4.8}{6} = 0.8$$

$$-e^{-t/RC} = -0.2$$

$$-t/RC = \ln 0.2 = -1.6$$

$$t = 1.6 \, RC$$

The delay is 1.6 time constants.

Given that R = 1 MΩ and C = 47 μF, the time constant is 47, and the delay is 1.6 × 47 = 75 s.

4　Describe a monostable based on the 7555 IC. It produces a pulse about 5 s long and can be reset by pressing a push-button.

5　Design an astable multivibrator based on a 555 IC, with a frequency of 10 Hz, and a mark-space ratio of 1.4.

6　Explain what is meant by the reactance and impedance of a capacitor.

7　Find f_C when $R = 2.2$ kΩ and $C = 470$ μF.

8　A highpass filter has its cut-off point at 33 Hz. If the resistor is 2.2 kΩ, what is the value of the capacitor?

9　A highpass filter has $R = 470$ Ω and $C = 4.7$ μF. What is its cut-off frequency? What is the impedance of the capacitor at this frequency?

10　Describe in outline the way in which a lowpass resistance/capacitor filter works.

11　Sketch the frequency response curve of a highpass filter in which $R = 330$ Ω and $C = 2.2$ nF, and mark the value of f_C.

12　Design the circuit of an RC bandpass filter, that is, one that filters out low and high frequencies but passes frequencies in the middle range. The pass-band should run from 10 kHz to 100 kHz. Check your design with a simulated circuit or on a breadboard.

Fields

When one object interacts with another object with which it is not in physical contact, it is because of a **field**. Think of a field as something that produces a force at a distance. If one or both objects are free to move, they will do so as a result of the force of the field.

Gravitation is an everyday example of a field. The force of gravity acts at a distance; for example, the Moon's gravity raises the ocean tides on Earth.

Two other kinds of field are important in everyday life — **electric fields** and **magnetic fields**.

ELECTRIC FIELDS

An electrically charged object is surrounded by an electric field. The charge may be positive or negative. The field affects other charged objects. If both have the same charge, they are repelled from each other and may move apart. If the two objects have opposite charges, they are attracted to each other.

Polarity

The terms *positive* and *negative* mean only that the two kinds of field are opposite. They do not mean that the positive field has something that is lacking from the negative field.

MAGNETIC FIELDS

A charged object produces an electric field whether it is moving or stationary but, when it *moves*, it produces an additional field, a **magnetic field**.

The magnetic field around a bar magnet, for example, is the result of the spinning and orbiting of the charged electrons in its atoms. Each atom is equivalent to a tiny magnet, each one having a north and south pole. The atoms may be aligned in random directions, so their fields cancel out for the bar as a whole.

But, if a sufficiently high proportion of them are aligned in the same direction, their fields combine and the material is said to be **magnetised**.

Magnetic flux, or **lines of force**, run from the north pole of the magnet to the south pole of the same or another magnet. They show the direction in which a compass needle aligns itself in the magnetic field.

Magnetic Poles

An object can have a single type of electric charge, positive *or* negative, but magnetic poles do not exist separately. Magnetised objects always have a **north pole** *and* a **south pole**.

Flux

Means 'flow'. The flux in a wire of solder helps the solder to flow when melted.

In the same way that two electric fields may interact to produce attractive or repulsive forces, two unlike poles attract each other, while two like poles repel each other.

FIELDS AND MOTION

When a charged object moves in a magnetic field it is subject to a force. The direction of the force depends on Fleming's **left-hand rule**.

The Left-Hand Rule

Hold your thu**M**b, **F**irst finger and se**C**ond finger at right-angles. Point your **F**irst finger in the direction of the electric **F**ield and your se**C**ond finger in the direction of the **C**urrent. Your thu**M**b points in the direction of the **M**otion.

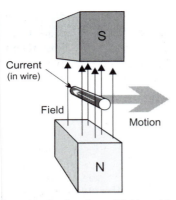

FIGURE 6.1 Illustrating the directions of fields and force in Fleming's left-hand rule (see previous page).

For example, if a current of electrons is moving along in a wire and the wire is in a magnetic field, each electron experiences a force. The resulting force on all the electrons appears as a force acting on the wire. Practical applications of this effect include electric motors and loudspeakers.

When a force moves a charged object in a magnetic field, it generates an electric field (an emf). The direction of the current obeys Fleming's **right-hand rule**.

For example, when a coil of wire is moved in a magnetic field, the emf generated causes an electric current to flow in the coil. This is called **electromagnetic induction**.

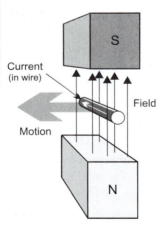

FIGURE 6.2 Fleming's right-hand rule for generating current.

The important point about the motion through the magnetic field is that the strength of the field through the coil *changes* as the coil is moved. We get the same effect if the coil stays in the same place and the field becomes stronger or weaker.

DESCRIBING ELECTRIC FIELDS

Electric field strength, *E*. An electric field acts on a charge Q with force F:

$$E = F/Q \text{ newton per coulomb}$$

This is numerically equal to the pd V between two points in the field d metres apart:

$$E = V/d \text{ volts per metre}$$

Electric field flux, φ. An electric field can be described by drawing lines that show the direction moved by charged objects within the field. These **lines of electric flux** are imaginary.

The electric field flux is proportional to the number of field lines passing at right-angles through a given area.

DESCRIBING MAGNETIC FIELDS

Magnetic flux density, B. A magnetic field produced by a unit charge moving at v m/s, in an electric field of E V/m has flux density:

$$B = v \times \frac{1}{c^2} E$$

c is the velocity of light in m/s, and equals 3×10^8 m/s. **B** is expressed in tesla (T).

Magnetic field strength, H. This is also expressed in tesla. **B** and **H** are related by the equation:

$$B = \mu H$$

In this equation, μ is the **magnetic permeability**, in henries per metre. The henry is described on p. 55. In a vacuum, $\mu = 1$, so **B** and **H** have the same value.

Force on a wire. When a current flows through a wire in a magnetic field, the force on the wire is:

$$F = BIL$$

F is the force in newtons, B is the flux density in teslas, I is the current in amps, and L is the length of the wire, in metres.

Magnetomotive force. Electromotive force (emf, p. 1) produces an electric flux or current. Magnetomotive force (mmf) produces magnetic flux. The unit of emf is the ampere-turn. When a direct and constant current of 1 A flows in a coil of 1 turn, the magnetic flux through the coil is 1 ampere-turn.

Note that the word *force* in the terms *electromotive force* and *electrocmagnetic force* does not mean that emf and mmf are mechanical forces. An electric

Topic | 6 Fields

cell, for example, produces emf as a result of chemical action, with no physical forces involved.

Magnetic reluctance: The lines of forces of a magnet are closed loops. Each loop is a magnetic circuit, equivalent to an electric circuit.

FIGURE 6.3 A single line of force produced by a bar magnet. The arrows indicate the direction of the field (the direction of a compass needle).

> ### On the Web
>
> There are several unexplained terms and devices mentioned in this topic. Examples: Hall effect sensors, series and shunt motors, and reed switches.
>
> For more information, search the Web using the terms or the device names as keywords.
>
> Use the Web to find out the properties and uses of ferromagnetic materials such as Ticonal, Alcomax, and Magnadur.

The reluctance of the magnetic circuit is equivalent to the resistance of an electric circuit. In an electric circuit or part of an electric circuit the resistance is R = emf/current. In a magnetic circuit the reluctance is:

$$S = mmf/magnetic flux$$

If there are objects with low reluctance in the field, the lines of force tend to crowd together and pass through the objects, avoiding the high-reluctance space between them. This is equivalent to the way in which electric current takes the route of least resistance.

When lines of force are concentrated through a low-reluctance material, the material develops north and south poles. It behaves as a magnet for as long as it is part of a magnetic circuit.

This explains the action of a solenoid. The plunger is made of soft iron, which has low reluctance, and it is not permanently magnetised. When the coil is energised, its magnetic field becomes concentrated through the plunger. The plunger becomes strongly

magnetised. The plunger is strongly attracted toward a region of high magnetic flux, so as to crowd even more lines of force into it. The result is that the plunger is *pulled* into the coil. A reed switch operates on a similar principle.

FERROMAGNETISM

This is the commonest type of magnetism. Bar magnets, compass needles, and the coating on magnetic hard disks all have ferromagnetism. It is a property of iron, cobalt, and compounds of these and other metals such as nickel, chromium and magnesium.

These materials have ferromagnetic properties because the spins of the electrons cancel out much less than in other materials. Groups of them align readily with an external magnetic field, forming magnetic domains. They align permanently so that the material becomes magnetised, and remains so, even when the external field is removed.

A piece of ferromagnetic material is magnetised by placing it in a magnetic field. The stronger the field, the more domains align themselves with it and the more strongly the material become magnetised. This is the basis of the hard disk drive of a computer. The relationship between the strength of the external field (H) and the magnetic flux density in the material (B) is illustrated in the graph below.

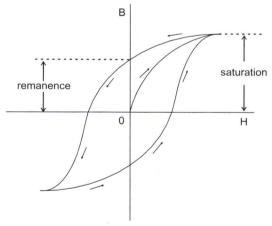

FIGURE 6.4 The effect of a magnetic field (*H*) on the magnetisation (*B*) of ferromagnetic material.

The diagram represents the effect on B of a sequence of changes in H. Follow the arrows. We begin at the origin (0) with unmagnetised material (B = 0) and no external field (H = 0).

As the external field is increased, B increases too, but the graph is not a straight line. It eventually levels out and further increase of H does not increase B. The material is **saturated**.

If H is gradually reduced, B falls too. But when H is reduced to zero, B is still greater than zero. The material has become magnetised.

Some of the magnetic field has remained (that is, some of the domains are permanently aligned). This remaining field is the remanence.

If the field is then reversed and increased in the reverse direction, B falls and eventually becomes saturated in the reverse direction. From that state, returning the field through zero and into the original direction will restore the original saturation. The effect is that changes in B *lag behind* changes in H. This lagging behind is called **hysteresis**.

Reversing the direction of magnetisation requires energy, the amount of energy being proportional to the area within the hysteresis curve (or loop). Magnetic materials differ in the amount of energy needed to reverse the magnetisation, as shown in Figure 6.5.

Hard magnetic materials need a strong field to magnetise them but have high remanence, so they make strong magnets. Once magnetised, they are not readily demagnetised by stray magnetic fields and so

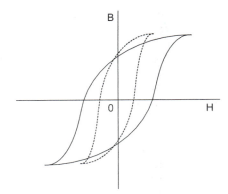

FIGURE 6.5   Typical B-H hysteresis loops of hard (continuous lines) and soft (dotted lines) magnetic materials.

are suitable for permanent magnets. Examples of such materials are ferrites, alloys such as Alcomax and neodymium-iron-boron (for the strongest magnets) and ceramic materials such as Magnadur.

Soft magnetic materials can be strongly magnetised by an external field but it takes little energy to demagnetise them or reverse their state. Examples are soft iron (used in electromagnets), cobalt alloys and chromium dioxide (used in magnetic hard disks).

Inductors

When a current is passed through a coil of wire, a magnetic field is generated. This is the principle on which solenoids, electric motors and loudspeakers work.

The converse happens too. When there is a change in the magnetic field through a coil, an emf is generated in the coil. This causes current to flow through the coil. This is known as **electromagnetic induction**.

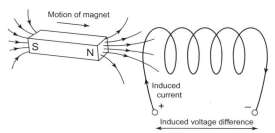

Motion of magnet

Induced
current
+   −

Induced voltage difference

FIGURE 7.1

There are three important points to note:

- **Emf is induced** in the coil only when the magnetic field is *changing* (Faraday's First Law). Current flows only if we move the magnet toward or away from the coil, or move the coil toward or away from the magnet. If both are still, there is no emf and no current.
- The **size** of the emf depends on the *rate of change* of the magnetic field (Faraday's Second Law). The faster we move the magnet or coil, the greater the emf and current.
- The **direction** of the induced current is such as to *oppose* the motion of the magnet or coil (Lenz's Law). In the drawing above, the north pole of the magnet is moving toward the coil. The direction of the current produces a north pole at the end of the coil that is nearer the magnet. Like poles repel, so the effect is to try to prevent the magnet from coming nearer. In practice, it does not prevent the motion but means that a little extra force is needed to move the magnet toward the coil. The extra energy

reappears as the energy of the current made to flow in the coil.

SELF INDUCTION

When the current through a coil changes, the magnetic field through the coil changes. This changing field acts just like a magnet being moved around near the coil — it induces another current in the coil. The direction of the current opposes the change in current through the coil. This effect, in which a coil induces current *in itself*, is called **self induction**.

The size of the induced current depends on the number of turns in the coil and other factors. The symbol for inductance and self inductance is **L**, and its unit is the **henry**, symbol **H**. Inductance can be defined in various ways, the most usual of which is:

$$E = -L \bullet \frac{\mathrm{d}I}{\mathrm{d}t}$$

E is numerically equal to the emf produced in a coil of self-inductance *L* henries when the current changes at the rate of 1 A per second. The negative sign indicates that the induced emf opposes the change of current.

> **Self Test**
>
> The current in a coil begins to fall at the rate of 6 A/s. If L = 840 mH, what emf is produced in the coil?

The same three points mentioned earlier apply also to self-induced currents:

- If there is *no change* in the current flowing through the coil, there is no induced current. DC flows through the coil without being opposed. It is affected only by the resistance of the wire of the coil.
- The size of the induced current is proportional to the *rate of change* of the magnetic field. If there is a very rapid change of current through an inductive component, such as when the current is switched off,

self induction produces a very large current which may damage other components in the circuit (see p. 16).

- The *direction* of the current is such as to oppose the changing current. If the current through the coil is increasing, the induced current acts to prevent the increase. Conversely, if the current is decreasing, the induced current acts to prevent it from decreasing. The result of this is that, if an alternating current is passed through the coil, its amplitude is reduced. This effect, which resembles resistance, is called the **reactance** of the inductor.

The inductance, and hence the reactance of a coil, are much increased if the coil is wound on a **core** of magnetic material. This acts to concentrate the lines of magnetic force within the core and so increase the strength of the magnetic field.

REACTANCE

The reactance of an inductor has the symbol x_L and is expressed in ohms. If a signal of frequency f is passed through an inductor of inductance L, its reactance is:

$$X_L = 2\pi f L$$

The equation shows that X_L increases with increasing f. At zero frequency (DC), X_L is zero. Direct current flows freely through an inductor coil. X_L is high at high frequency so radio-frequency signals are much reduced in amplitude when passed through an inductor coil.

Inductive reactance can also be expressed as inductive impedance, Z_L.

TYPES OF INDUCTOR

Many kinds of component have self-inductance. Solenoid coils, loudspeakers and other devices that include a coil are examples. Even the leads of a capacitor, resistor or transistor, though not coiled, have significant self-inductance at radio frequencies. Inductance in these cases may be ignorable, or may cause problems.

Inductance is a useful property of certain other components. These include:

Chokes: They are used to block high-frequency signals from passing through from one part of a circuit to another. Low-frequency signals or DC voltage levels are able to pass through. Large chokes look like transformers, but have only one coil. Small chokes consist of beads or collars made of ferrite,

threaded on to the wire that is carrying the high-frequency signals.

FIGURE 7.2 Ferrite beads suppress high-frequency signals passing along the conductor.

Ferrite is an iron-containing material, so it acts as a core to contain the lines of magnetic force around the wire. Sometimes a choke is made by winding the wire around a ferrite ring.

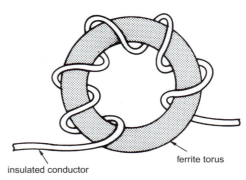

FIGURE 7.3 High-frequency signals in a conductor may be damped by winding the wire around a ferrite torus.

Tuning coils: Used in radio transmittters and receivers to tune a circuit to a particular radio frequency.

The coil is wound on a plastic former. It may have a core of ferrite or iron dust ceramic that can be screwed in or out of the coil to tune it.

When connected to a capacitor, it forms a tuned network that resonates at a particular frequency, as explained below.

FIGURE 7.4 This VHF tuning coil with 3.5 turns of wire has a self inductance of 0.114 μH.

The inductor may be a few turns of wire wound on a plastic **former,** as in Figure 7.4. A tuned inductor may have more than one winding, and is equivalent to a transformer that operates best at one particular frequency.

One or more coils are wound on a plastic former. The iron-dust core is threaded, as is the inside of the former. The core is moved into or out of the coil by turning it with a special non-magnetic tool so as not to introduce spurious magnetic effects. This allows the self inductance of the coil to be adjusted precisely. Inductors of this type are used in the same way as capacitors, for coupling two parts of a circuit (pp. 34−35). However, they are suited only to radio-frequency signals.

ENERGY TRANSFERS

Resistance is fundamentally different from capacitance and inductance, even though they are all forms of impedance. When a current passes through a resistance, some of its energy is converted irreversibly to other forms. In most cases it is converted to heat, though lamps and LEDs produce light and motors produce motion. The loss appears as a loss of potential across the resistance.

When current passes into a capacitor, the charge is stored on its plates. The stored energy appears as a difference of potential between the plates. The charge is released later when the capacitor is discharged. There is no loss of energy during charging and discharging. In theory at least, no heat or other form of energy is released.

Similarly, when the current through an inductor increases, electrical energy is stored in the form of a magnetic field. When the current decreases, the field wholly or partly collapses, releasing energy as the induced emf. Again, there is no loss of energy in this process. No heat or other form of energy is produced.

L-C NETWORK

The impedance of a capacitor decreases with increasing frequency. The impedance of an inductor is the opposite of this. Combining a capacitor and an inductor in the same network results in an interesting action. Figure 7.5 shows the circuit.

V_{IN} is a sinusoid, frequency f. At low frequencies the inductor has low impedance. The signal passes easily through it, so V_{OUT} is small. At high frequencies the capacitor has low impedance. The signal passes easily through it, so V_{OUT} is again small. There is one frequency, between high and low, at which the inductor and capacitor have equal

FIGURE 7.5

impedances. At this frequency the signal passes less easily through them and V_{OUT} is a maximum.

Calling this frequency f, the impedance of C is:

$$X_C = 1/2\pi f C$$

At the same frequency, the impedance of L is:

$$X_L = 2\pi f L$$

When $X_C = X_L$:

$$\frac{1}{2\pi f C} = 2\pi f L$$

$$f^2 = \frac{1}{4\pi^2 LC}$$

$$f = \frac{1}{2\pi\sqrt{LC}}$$

With the values of C and L shown in the diagram:

$$f = 1/(2\pi \times \sqrt{(1 \times 100)})\left[1/(mH \times nF)\right]$$

$$f = 15.9 \text{ kHz}$$

To confirm these calculations, the frequency response of the circuit is:

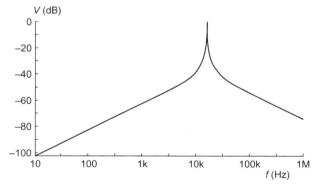

FIGURE 7.6

The frequency response peaks very sharply at 16 kHz. At this frequency the alternating signal has the maximum effect on the network, and the network is said to **resonate.**

reduced signal can eventually appear from the amplifier.

In (b) a follower amplifier is connected as a buffer between the microphone and the audio amplifier.

The follower amplifier has high input resistance R_I and low output resistance R_O. Because R_I is much greater than R_{OUT}, most of v_S appears across R_I and only a small part of it is 'lost' across R_{OUT}. We now have transferred most of the original signal to the follower amplifier. Here its voltage is not amplified, so we may still represent the signal by v_S.

Now the resistance through which the signal has to pass is R_O, which is very much *smaller* than R_L. This being so, only a small portion of v_S is 'lost' across R_O and most of it appears across R_L. Most of the signal from the microphone has been transferred through the follower amplifier to the audio amplifier, where it can be amplified further and perhaps fed to a speaker.

This technique for using a follower amplifier for transferring the maximum signal is known as **impedance matching**.

CURRENT GAIN AND POWER GAIN

Although the voltage gain of a MOSFET follower amplifier is only about 1, it has high current gain. The input current to the insulated gate is only a few picoamps, while the output current is measured in milliamps or even amps. So the current gain of MOSFET voltage amplifiers is very high. It is not possible to put an exact figure on the gain because the current entering the gate is a leakage current, and may vary widely from transistor to transistor.

Because current gain is high and power depends on current multiplied by voltage, the power gain of MOSFET amplifiers is very high.

ACTIVITY — MOSFET C-D AMPLIFIER

Using the circuit on p. 67 as a guide (Figure 8.6), build a single-stage MOSFET common-drain amplifier. Use the component values quoted in the diagram. Connect a signal generator to its input and an oscilloscope to its output. Observe the input and output signals. Find the gain of the amplifier at 1 kHz.

Next, try varying the frequency of the signal in steps over the range 10 Hz to 100 MHz. Keep the amplitude at 100 mV. Measure the output amplitude at each frequency.

Plot a graph of output against frequency. It is best to use a logarithmic scale for frequency, as on p. 66. There is no need to convert the output amplitudes to the decibel scale.

Try increasing the amplitude of the input signal and observe the amplitude of the output signal. Find out what is the largest input amplitude that this circuit can follow without distortion.

Try altering R1 and R2 to see the effect of biasing. For example, make $R_2 = 100\,k\Omega$. Then make $R_2 = 1\,M\Omega$.

MOSFET APPLICATIONS

MOSFETs are used in ON-OFF transistor switches (Topic 2) and in amplifiers (this topic). They are also used as analogue switches (Figure 8.8). The analogue signal passes (in either direction) when the control input (gate) is high.

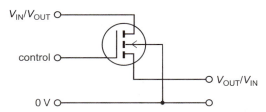

FIGURE 8.8 In the analogue switch, the substrate of the MOSFET is connected to 0 V.

A MOSFET can also act as a voltage-controlled resistor. The voltage output of an attentuator circuit varies with the control voltage. Attenuators are used to provide automatic gain control (AGC) in amplifiers. They are also used in amplitude modulation (AM) circuits.

FIGURE 8.9 A MOSFET attenuator works like a potential divider in which one of the resistors (Q1) is variable.

EXTENSION BOX 15 Types of MOSFET

In **n-channel** MOSFETs (Figure 8.10) the drain current is conducted as electrons through n-type material. The transistor is made of p-type semiconductor with isolated electrodes of n-type connected to the source and drain terminals. Conduction from source to drain can not occur because conduction from p-type to n-type at the drain electrode can not occur. This is because the p-n junction there acts as a reverse biased diode.

0 V
source

+V_g
gate

+V
drain

substrate
connection

induced
n-channel

→● free electron

FIGURE 8.10 The effect of the field at the gate is to create a channel of n-type silicon connecting the two existing n-type regions.

When the gate is at a positive potential, electrons are attracted toward the region of the gate. This creates a continuous zone of n-type electrons joining the two electrodes. Now the conduction path is completely n-type, as shown in the diagram, and flow of electrons from source to drain can occur.

The greater the potential of the gate, the more electrons are attracted and the lower the resistance to the flow of drain current.

In **p-channel** MOSFETs, the current is conducted as holes through p-type material. The transistor is made of n-type material, with p-type electrodes. A negative potential at the gate repels electrons from that region. This creates a continuous region of p-type material joining the electrodes and a 'current' of holes can flow.

In either type of MOSFET there is only one type of charge carrier (electrons or holes), which gives these transistors the description **unipolar**. Both n-type and p-type MOSFETs work by the creation of a channel when a suitable potential is applied to the gate. These MOSFETs are therefore described as **enhancement** MOSFETs.

It is possible to manufacture n-channel MOSFETs which have n-type substrate. A conductive channel is present in these when the gate and source are at the same potential. The channel is *reduced* in width as the gate is made more negative of the source and electrons are repelled from its vicinity. These transistors are known as **depletion** MOSFETs. They are rarely used and are not commonly listed in suppliers' catalogues.

QUESTIONS ON MOSFET AMPLIFIERS

1 Explain what is meant by transconductance. How is the output current of a MOSFET converted to an output voltage in a common-source amplifier?

2 Why is it important to keep signal amplitudes small in MOSFET amplifiers?

3 Why is it usually preferable to use capacitor coupling at the input and output of an amplifier, rather than use wired connections?

4 Explain how a common-drain amplifier is used to match a signal source with a high output resistance to an amplifier with low input resistance.

5 Explain what the −3 dB point means.

6 What is the advantage of the high input resistance of MOSFET amplifiers? Give two practical examples of this.

7 Why do we usually try to make the quiescent output voltage of an amplifier equal to half the value of the supply voltage?

8 Outline the practical steps in testing the frequency response of an amplifier. Why are the results usually plotted on logarithmic scales?

9 Explain why the amplitude of an amplifier usually falls off at low and high frequencies.

10 Design an amplifier using an n-channel MOSFET with operating voltage 10 V, and able to amplify signals of 200 Hz and over.

11 The supply voltage of a common-drain MOSFET amplifier is 9 V. The gate is biased to 4 V by a pair of resistors. The amplifier is to operate at frequencies of 100 Hz and over. Calculate suitable values for the input coupling capacitor and resistors.

MULTIPLE CHOICE QUESTIONS

1 A MOSFET is described as a unipolar transistor because:

A current flows through it in only one direction

B it may explode when connected the wrong way round

C it has only one type of charge carrier

D the channel is made of n-type material.

2 A follower amplifier is often used as a buffer between:
A a high resistance output and a low resistance input
B a low resistance output and a high resistance input
C a high resistance output and a high resistance input
D a low resistance output and a low resistance input.

3 A follower amplifier has a gain of:
A approximately 100
B exactly 1
C slightly less than 1
D slightly more than 1.

4 The total parallel resistance of resistors of 56 kΩ and 120 kΩ is:
A 38.2 kΩ
B 176 kΩ
C 64 kΩ
D 6720 kΩ.

5 The advantage of using a capacitor to couple an amplifier to a signal source is that:
A it stabilises the gain
B the amplifier is not affected by the quiescent output voltage of the source
C the capacitor acts as a highpass filter
D the capacitor filters out noise from the signal.

6 In p-channel MOSFETs the charge carriers flow from:
A drain to source
B source to gate
C source to drain
D negative to positive

7 For a MOSFET to conduct, its gate-source voltage must be:
A negative
B above 3 V

C greater than the threshold
D 0.7 V.

8 A common-source amplifier:
A is an inverting amplifier
B has unity gain
C has high output resistance
D has low current gain.

9 The function of the drain resistor of a common-drain amplifier is to:
A limit the output current
B amplify the signal
C convert the drain current to a voltage
D act as a lowpass filter.

10 The resistor and capacitor at the output of a common-drain amplifier act as a:
A potential divider
B lowpass filter
C current to voltage converter
D highpass filter.

EXTENSION QUESTIONS

1 Describe the structure and action of an n-channel enhancement MOSFET.

2 Design a common-source amplifier based on a p-channel enhancement MOSFET. Component values are not required.

3 Describe how conduction occurs in (a) n-type and (b) p-type semiconductors.

4 Why is there a forward voltage drop acoss a p-n junction?

BJT Amplifiers

There are three basic ways in which a bipolar junction transistor may be used as an amplifier: common-emitter amplifier, common-collector amplifier and common-base amplifier.

These each have very different attributes and applications. We look at the circuits of the first two of these and also a few other types of BJT amplifier.

COMMON-EMITTER AMPLIFIER

The BJT amplifies current. In the simple common-emitter amplifier below, based on an npn BJT, a small current i_B flows through R1 into the base of Q1. Because of transistor action, this causes a much larger current i_C to flow in at the collector of Q1. The currents combine and flow out of Q1 at the emitter.

FIGURE 9.1 In a common-emitter BJT amplifier the emitter terminal is common to both the input and output circuits. Compare with the MOSFET common-source amplifier on p. 63.

The emitter current i_E is:

$$i_E = i_B + i_C$$

These currents are shown on the circuit diagram above.

CURRENT GAIN

The relationship between i_B and i_C is known as the **forward transfer characteristic**. This graph is plotted by using a test circuit like that in Figure 9.3. Compare this with the test circuit on p. 64, used for measuring the transconductance of MOSFETs.

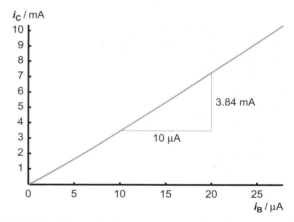

FIGURE 9.2 The graph of output current against input current of a BJT is almost a straight line. Compare this with the graph on p. 64, where the gradient of the curve increases with increasing input voltage.

This one has two current meters, because both the input and output of a BJT are currents. Note that the meter for measuring i_B is a microammeter, while the meter for i_C is a milliammeter. This is because i_C is always much larger than i_B.

We can measure how much larger by plotting the graph of i_C against i_B. This shows that, if i_B is increased by 10 μA, i_C increases by 3.84 mA.

The ratio between the two (or the slope of the curve) is known as the **small signal current gain**, symbol h_{fe}. In this example:

$$h_{fe} = 3.84/10 \ [mA/μA] = 384$$

Electronics: Circuits and Systems.

FIGURE 9.3   This is the kind of circuit used for measuring the forward transfer characteristic of a BJT.

The gain for the BC548 is often listed as 400 (but see below). Note that the graph is almost a perfect straight line.

Self Test

What is the h_{fe} of a BJT if i_C increases by 2.06 mA when i_B is increased by 5 μA?

Because the curve is almost a straight line we can measure gain in a slightly simpler way. Reading the value of i_C on the milliammeter when $i_B = 20$ μA, we find $i_C = 7.68$ mA. Now we calculate the **large signal current gain**, symbol h_{FE}:

$$h_{FE} = 7.68/20 \text{ [mA/μA]} = 384$$

As is expected from the geometry of a straight-line curve, the gains are equal. The difference between h_{fe} and h_{FE} is theoretical, and rarely practical.

VOLTAGE OUTPUT

In the CE amplifier, the function of R2 is to convert the current i_C into a voltage, according to Ohm's Law. The values give a quiescent collector current i_C of 1 mA, which is a typical value for a CE amplifier. R2 is chosen so that the voltage drop across it is equal to half the supply voltage at this current. This allows the output voltage to rise and fall freely on either side of this half-way voltage without distortion.

Example

Current = 1 mA, voltage drop = 7.5 V, so:

$$R_1 = 7.5/0.001 = 7500$$

R_1 should be 7.5 kΩ.

BIASING

A suitable size for the collector current in a common-emitter amplifier is 1 mA. This minimises noise (p. 229). The value of R1 is often chosen to provide a base current sufficient to produce such a collector current.

Example

Current gain = 380, and $i_C = 1$ mA, so:

$$i_B = 1/380 \text{ [mA]} = 2.63 \text{ μA}$$

Assuming that the base-emitter voltage drop v_{BE} is 0.7 V, the voltage drop across R1 is $15 - 0.7 = 14.3$ V. The resistance of R1 is:

$$R_1 = 14.3/2.63 \text{ [V/μA]} = 5.44 \text{ MΩ}$$

We could use a 5.6 MΩ resistor. But to allow for the gain of the transistor to be a little less than 380, make i_B a little larger by making R1 a little less. We decide on 4.7 MΩ.

Base-Emitter Voltage

v_{BE} is usually between 0.6 V and 0.7 V for a silicon transistor.

This illustrates one of the problems with such a simple circuit. The correct value of R1 depends on the gain of the transistor. But transistors of the same type vary widely in their gain. Any given BC548 may have h_{fe} between 110 and 800. When building this circuit it is almost essential to match R1 to the particular transistor you are using.

EMITTER RESISTANCE

From the circuit diagram, it looks as if this circuit has a very high input resistance, the resistance of R1. But, unlike the gate of a MOSFET, the base of a BJT

FIGURE 9.4   When current flows from the base of a BJT to the emitter, it is subject to the emitter resistance.

is *not* insulated from the body of the transistor. Current i_B flows into the base, through the emitter layer and out by the emitter terminal. Along this path it encounters resistance, typically about 25 Ω. This is the emitter resistance r_e, which can be thought of as a resistor inside the transistor.

In the way it affects current entering the base, this resistance *appears* to be h_{fe} times its actual value.

FREQUENCY RESPONSE

As with the MOSFET amplifiers, C1 and R1 form a highpass filter. The value of C1 is chosen to pass all signals above 20 Hz. The values of R2 and C2 are chosen to pass the same frequencies on the output side.

VOLTAGE GAIN

We will work through a rough calculation to find the voltage gain of the amplifier.

The voltage gain of this amplifier is:

$$A_v = -3/10 \ [\text{V/mV}] = -300$$

The gain is 300, and the negative sign indicates that the signal is inverted.

There are several approximations in this calculation but the result agrees well with a practical test, which showed a voltage gain of −250.

IMPROVING STABILITY

The main disadvantage of the amplifier shown in p. 71 is its lack of stability. Its performance depends on the value of h_{fe}, which varies from transistor to transistor. Also, h_{fe} is affected by temperature. The amplifier may be operating in a particularly hot or cold environment which would make its performance even more variable. Even in a room at a comfortable temperature, the temperature of the amplifier increases after it has been run for a while.

The circuit is improved if we connect the positive end of R1 to the collector of Q1. The positive end of R1 is now held at *half* the supply voltage, so we halve its resistance to obtain the same i_B as before. This does not affect the amplification but has an effect on stability.

FIGURE 9.5 Stability is improved when R1 connects the collector to the base.

To see how this works, suppose that for some reason h_{fe} is higher than was allowed for in the design. Perhaps the transistor has a higher than average h_{fe} or perhaps h_{fe} has been raised by temperature. If h_{fe} is higher than usual, i_C is *larger* than usual. If i_C is larger, the voltage drop across R2 is greater. This lowers the voltage at the collector and so lowers the voltage at the positive end of R1. This reduces i_B, which in turn *reduces* i_C.

In summary, a change which increases i_C is countered by a reduction in i_C. The result is that i_C tends to remain unaffected by variations in h_{fe}. The gain of the amplifier is stable.

Further improvement in stability is obtained by biasing the base with two resistors, instead of only one:

FIGURE 9.6 Biasing the base with a voltage divider helps stabilise the common-emitter amplifier.

R1 and R2 act as a potential divider to provide a fixed quiescent voltage, independent of h_{fe}. We have also introduced an additional improvement, the **emitter resistor**, R4. This is not to be confused with the emitter *resistance*, discussed on p. 72.

The emitter resistance and emitter resistor are in series but the emitter resistance is usually about 25 Ω and may be ignored when the emitter resistor has a value of 1 kΩ or more. If i_C is 1 mA, a convenient value for R4 is 1 kΩ, which produces a voltage drop of 1 V in the quiescent state. Now that the emitter voltage is 1 V, the base voltage must be 0.7 V more than this, to maintain the required v_{BE} of 0.7 V. The values of R1 and R2 are calculated so as to produce 1.7 V at the base.

INPUT RESISTANCE

The input resistance of the amplifier is the total of three resistances in parallel: R1, R2, and R4.

As with emitter resistance, a current flowing into the base is affected by h_{fe} times the resistance between the base and the 0V line. Now that there is a 1 kΩ resistor there, we can ignore r_e. The resistance of R4, as seen from the base, is h_{fe} times 1 kΩ, that is, 400 kΩ. The total of these three resistances in parallel is 27 kΩ. This gives the amplifier a moderate input resistance, but one which is far less than that of MOSFET amplifiers.

Given the new value of the input resistance, we can calculate a new value for C1.

Self Test

Find a value for C1 which will filter out frequencies below 1 MHz.

VOLTAGE GAIN

In this discussion, v_{in} means 'a small change in v_{IN}', and the same with other symbols.

In the amplifier on the left the signal passes more-or-less unchanged across C1:

$$v_{in} = v_b$$

where v_b is the base voltage. This stays a constant 0.7 V higher than v_E, the emitter voltage. We can write:

$$v_{in} = v_b = v_e$$

Applying Ohm's Law, we also know that:

$$i_e = v_e/R_4 \quad \text{and} \quad v_{out} = -i_c \bullet R_3$$

The negative sign indicates that we are dealing with a voltage *drop* across R_3.

From these equations we find that:

$$v_{out} = -i_c \bullet R_3 = -i_e \bullet R_3 = -v_e \bullet R_3/R_4 = -v_{in} \bullet R_3/R_4$$

From this we obtain:

$$\text{voltage gain} = v_{out}/v_{in} = -R_3/R_4$$

The result shows that the voltage gain depends only on the values of the collector and emitter resistors. The voltage gain is independent of h_{fe}, so *any* transistor, either of the same type or of a different type, will produce the same voltage gain. Voltage gain is also virtually independent of temperature. The amplifier has high stability, but at the expense of low gain.

Example

In the amplifier with $R_3 = 7.5 \text{ k}\Omega$ and $R_4 = 1 \text{ k}\Omega$, the expected voltage gain is:

$$-7.5/1[\text{k}\Omega/\text{k}\Omega] = -7.5$$

A test on this circuit showed an actual voltage gain of -7.25.

FREQUENCY RESPONSE

With the capacitor values given in Figure 9.6, the frequency response of the circuit shows the lower cut-off point to be at 20 Hz, as calculated. In the higher frequencies, the cut-off point is approximately 11.7 MHz. This bandwidth of 11.7 MHz is satisfactory for many applications.

An upper cut-off point of 11.7 MHz may be essential for an amplifier in a radio receiver, but it is unnecessary for an audio amplifier. The human ear is not sensitive to frequencies higher than about 20 kHz. In Topic 27, we show that there are disdvantages in amplifying frequencies outside the required range.

BYPASS CAPACITOR

The use of an emitter resistor (R4) gives improved stability but reduced gain. If required, the gain can be restored by wiring a high-value capacitor across R4. This holds the emitter voltage more or less constant, so the calculations on p. 74 do not apply. Without the capacitor, the voltage at the emitter rises and falls with the signal. This provides negative feedback.

FIGURE 9.7 The addition of a bypass capacitor holds the emitter voltage steady and maintains voltage gain.

For example as i_B rises (tending to increase v_{BE}), i_C rises, and the emitter voltage rises. This tends to decrease v_{BE}, which decreases i_C and counters the rise in emitter voltage. Another way of looking at this is to say that the capacitor shunts the signal at the emitter through to the ground. For this reason C3 is called a **bypass capacitor**. With this capacitor in place, the voltage gain of the amplifier is about 280. The lower cut-off point is raised to 130 Hz, so bandwidth is slightly reduced.

POWER

Common-emitter amplifiers have current flowing through the transistor even when no signal is being amplified. This type of amplifier is known as a **Class A amplifier**. Power is being wasted when there is no signal, which makes Class A amplifiers unsuitable for

high-power amplification. We discuss power amplifiers in Topic 15.

ACTIVITIES — BJT AMPLIFIER

1 To measure the forward transfer characteristic of a BJT, set up the circuit illustrated on p. 72. V_{CC} is a fixed voltage source of 15 V DC, v_{BE} is a variable voltage source. Adjust the v_{BE} to produce a range of base currents, from zero to 25 μA in steps of 5 μA. Measure the corresponding values of i_C. Plot i_C against i_B and calculate the small signal current gain.
2 Investigate the action of a BJT common-emitter amplifier. Set up the circuit on p. 74. Connect a signal generator to its input and an oscilloscope to its output. Using a 100 mV sinusoidal signal, find the voltage gain of the amplifier at 1 kHz and other frequencies.
3 Vary the values of R3 and R4 to confirm that these determine the voltage gain of the amplifier. Try this with several different transistors of known h_{fe} to show that voltage gain is independent of h_{fe}.
4 Investigate the effect of the bypass capacitor.

COMMON-COLLECTOR AMPLIFIER

The common-collector amplifier below has an emitter resistor but the collector is connected directly to the positive rail. The base is biased by two resistors. Assuming that the quiescent emitter current is 1 mA, the voltage across the emitter resistor R3 is 7.5 V, bringing the output to exactly half-way between the supply rails. To provide for a v_{BE} of 0.7 V, we need to hold the base at 8.2 V. The values of R1 and R2 are calculated to provide this.

FIGURE 9.8 The voltage output of a common-collector amplifier is always approximately 0.7 V lower than its input.

VOLTAGE GAIN

We calculate the voltage gain of the amplifier by considering what happens when the input voltage v_{IN} changes by a small amount v_{in}. The change in v_{IN} is carried across C1 to the base of Q1.

The change in base voltage is given by:

$$v_b = v_{in}$$

As the base voltage changes, the drop (v_{BE}) across the base-emitter junction remains constant. Therefore, the emitter voltage v_E changes by the same amount, which is v_b. This change is transferred across C2 to the output terminal.

For all such small changes in voltage we can say that:

$$v_{out} = v_e = v_b = v_{in}$$
$$\text{voltage gain} = v_{out}/v_{in} = 1$$

A gain of 1 is often referred to as **unity gain**.

The amplifier is a non-inverting amplifier with unity voltage gain. The output exactly follows all *changes* in the input, but is 0.7 V lower because of the drop across the base-emitter junction. If the signal is large enough, this drop can be ignored. Because the output follows the input, we refer to this kind of amplifier as a **voltage follower**. It is also known as an **emitter follower**.

The input resistance is high, being equal to the total resistance R_1, R_2, and $h_{fe} \times R_3$. With the values shown, the input resistance is 70 kΩ. The output resistance is the value of R3, which in this case is 7.5 kΩ. The output resistance is only about one tenth of the input resistance, making the amplifier suitable as a buffer (see pp. 67–68).

CURRENT AND POWER GAIN

Because input current is small (due to the high input resistance) and output current is moderately large, this amplifier has high current gain and high power gain.

FREQUENCY RESPONSE

The frequency response is very good. In the case of the common collector amplifier in Figure 9.8, the value of C1 is chosen to produce a highpass filter with lower cut-off point at 20 Hz. The upper cut-off point is at 128 GHz. The reason for this high cut-off point is that a transistor with its collector connected directly to the positive rail is not as subject to the effects of capacitance as one which is connected through a resistor. A common-collector amplifier can be said to have a wide bandwidth.

DARLINGTON PAIR

A Darlington pair consists of two BJTs connected as below. The emitter current of Q1 becomes the base current of Q2. The current gain of the pair is equal to the product of the current gains of the individual transistors.

Example

If $h_{fe} = 100$ for each transistor in a Darlington pair, the gain of the pair is $100 \times 100 = 10\,000$.

FIGURE 9.9 A Darlington pair has very high current gain.

A Darlington pair may be made up by wiring two individual transistors together. The two transistors may also be manufactured as a unit on a single silicon chip. This has three terminals, collector, base and emitter. The chip is enclosed in a normal three-wired package looking just like an ordinary transistor. When the pair is made from individual transistors, they need not be of the same type. Very often Q1 is a low-voltage low-current type while Q2 is a power transistor.

Self Test

A Darlington pair is made up of a BC548 with $h_{fe} = 400$ and a TIP31A (a power transistor) with $h_{fe} = 25$. What is the gain of the pair?

A Darlington may replace the single transistor used by the BJT common-emitter and common-collector amplifiers. This gives increased sensitivity to small base currents and, in the case of the common-emitter amplifier, may result in increased gain.

Darlingtons may also be used as switching transistors, as in Topic 2. They require only a very small current to trigger them.

The circuit below is a practical example of a Darlington pair being used as a switching transistor.

FIGURE 9.10 Using a Darlington pair in a touch-switch to operate a buzzer.

Two metal plates about 1 cm × 2 cm are mounted side by side with a 0.5 mm gap between them. When a finger touches the plates so as to bridge the gap, a minute current flows to the base of Q1. This current is only a few tens of microamperes, but the pair of transistors has a current gain of about 160 000. The resulting collector current through Q2 is sufficient to power the solid-state buzzer.

ACTIVITY — DARLINGTON SWITCH

Set up a Darlington pair as a switch controlling an audible warning device (see diagram above). A pair of BC548 transistors are suitable. Alternatively, use a Darlington pair transistor such as an MPSA13.

When the circuit is working, insert a microammeter and a milliammeter into it to measure the current through the finger and the current through the AWD.

What is the current gain of your circuit?

DIFFERENTIAL AMPLIFIER

This differential amplifier (below) is also known as a **long-tailed pair**. A differential amplifier has two inputs v_{IN+} and v_{IN-}. The purpose of the amplifier is to amplify the voltage *difference* between its inputs.

When it does this, we say that it is operating in the **differential mode**.

FIGURE 9.11 A differential amplifier has two inputs and one or two outputs. The second output (not shown here) is taken from the collector of Q1.

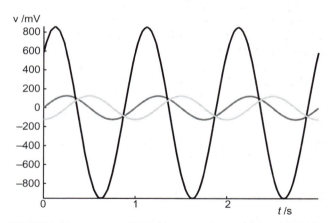

FIGURE 9.12 With the differential amplifier in differential mode, the output (black curve) is proportional to the difference between the non-inverting input (dark grey curve) and the inverting input (light grey curve). The inputs have 5 mV amplitude and are plotted on a 25 times scale.

To demonstrate this, two sinusoid signals that are 90° out of phase are fed to the two input terminals. The output signal is taken from the collector of Q2. At any instant:

$$v_{OUT} = (v_{IN+} - v_{IN-}) \times \text{differential-mode voltage gain}$$

Measurements on the graph (above) show that the differential voltage gain is approximately 129.

If we feed the *same* signal to both inputs, we are operating the amplifier in **common-mode**. We obtain the result shown in Figure 9.13. The amplitude of the output signal is only half that of the input signal and it is inverted:

$$v_{OUT} = v_{IN} \times \text{common-mode gain}$$

The common-mode gain is -0.5.

The differential amplifier has high differential-mode gain but very low common-mode gain. It is suited to measuring voltage differences when the voltages themselves are subject to much larger changes affecting both voltages equally.

In medical applications we may need to measure very small differences of voltage between two probes attached to different parts of the body surface. At the same time both probes are subject to relatively large voltage changes resulting from electromagnetic fields from nearby mains supply cables and from other electrical equipment in the same room.

The differential amplifier is able to amplify the small voltage differences between probes but is much less affected by the larger voltage swings affecting both probes. It is able to *reject* the common-mode signals. We express its ability to do this by calculating the **common-mode rejection ratio**:

$$\text{CMRR} = \frac{\text{diffl. mode voltage gain}}{\text{common mode voltage gain}}$$

Example

We use the differential mode voltage gain as taken from the graph. We ignore negative signs when calculating the CMRR:

The CMRR is $129/0.5 = 258$.

The CMRR can also be expressed in decibels:

$$\text{CMRR} = 20 \times \log_{10}(129/0.5) = 48.2 \text{ dB}$$

The common mode rejection ratio is 258, or 48.2 dB.

ACTIVITIES — DIFFERENTIAL AMPLIFIER

Build the differential amplifier (Fig. 9.1) on a breadboard or simulate it on a computer.

(a) First measure the **differential-mode voltage gain**. Connect a voltage source to each of the inputs and a voltmeter to the output. Draw a table with 5 columns. Head the first 3 columns v_{IN+}, v_{IN-} and v_{OUT}. Set the two voltage sources to two different voltages within the range ± 100 mV.

Record the two voltages (v_{IN+} and v_{IN-}). Measure and record v_{OUT}. Repeat this for 10 pairs of input voltages. For each pair calculate ($v_{IN+} - v_{IN-}$) and write this in the fourth column. For each pair calculate the differential-mode voltage gain and write this

FIGURE 9.13 A simulation of the differential amplifier in common mode. The output (black curve) has smaller amplitude than the input (grey curve) and is inverted, demonstrating a gain of -0.5.

in the fifth column. Calculate the average value of this for the 10 pairs.

(b) Next measure the **common-mode voltage gain**. Connect the two inputs together and connect a single voltage source to them. Draw a table with three columns and in the first two columns record v_{IN} and v_{OUT} as you obtain 10 sets of readings with the input in the $\pm 100\,\text{mV}$ range. For each set of readings, calculate the common-mode voltage gain and write this in the third column. Calculate the average value for this for the 10 sets.

(c) Finally, calculate the **common-mode rejection ratio**. Using the two gains found in (a) and (b), calculate the common-mode rejection ratio. Express this in decibels.

TUNED AMPLIFIER

Tuned amplifiers are used at radio frequencies. Their essential feature is a tuned network, usually consisting of a capacitor in parallel with an inductor (p. 57). There are several ways in which the LC network may be built into the amplifier, but one of the commonest ways is to use it to replace the resistor across which the voltage output signal is generated. In a BJT common-emitter amplifier it replaces the collector resistor. In a MOSFET common-source amplifier, it replaces the drain resistor. Below is a basic common-emitter amplifier (Figure 9.8, p. 75) with its collector resistor R_3 replaced by an LC network. The transistor has been replaced by a high-frequency type.

The resonant frequency of the network is given by:

$$f_0 = \frac{1}{2\pi\sqrt{LC}}$$

FIGURE 9.14 A tuned amplifier based on a simple common-emitter amplifier.

Example

The resonant frequency of the LC network is:

$$f_0 = \frac{1}{2\times\pi\times\sqrt{3.3\times820}} = 96.75\,\text{kHz}$$

$$[1/(\text{mH}\times\text{pF})]$$

The LC network is tuned to resonate at 96.75 kHz. If the amplifier is fed with a 10 mV signal in a range of frequencies from 95 kHz to 100 kHz, we can obtain the frequency response plot shown below. Looking at the curve nearest the front of the diagram, it can be seen that the output amplitude peaks sharply at just under 97 kHz, as predicted by the calculation above.

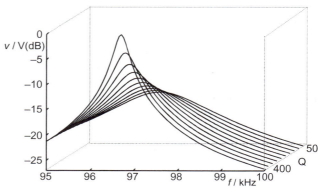

FIGURE 9.15 The frequency response of the tuned amplifier varies according to the Q of the inductor.

The reason for this peak is the fact that, at the resonant frequency, the capacitor-inductor network behaves as a pure resistance of high value. This is equivalent to replacing the network with a resistor equivalent to the one that was present as R3 in the amplifier on p. 74.

As the collector current flows through the network, a voltage is developed across it, and this is the output of the amplifier. At lower or higher frequencies the effective resistance of the network is not as great, either the capacitor or the inductor having low resistance. The voltage developed across it is less at lower or higher frequencies.

Because the resonating network has a high resistance (in theory, an infinite resistance, but less than this in practice due to losses in the capacitor and inductor) the voltage developed across it is high. In the plot above, the amplitude of the output signal at the resonant frequency is 15 V. This is a voltage gain of 1 500 times.

6 Describe the action and applications of a common-collector amplifier.

7 Compare the common-emitter BJT amplifier with the common-source MOSFET amplifier.

8 Explain what is meant by negative feedback. Quote two examples of negative feedback in amplifiers.

9 A transistor has an h_{fe} of 110. By how much does the collector current increase when the base current is increased by 25 µA?

10 Describe a Darlington pair and state how to calculate its current gain.

11 What are the advantages of having a Darlington transistor as a single package?

12 Why may we want to build a Darlington transistor from two separate BJTs?

13 Design a rain-alarm circuit for washday, using a Darlington transistor.

14 Draw the circuit of a BJT differential amplifier and state the relationship between its inputs and output.

15 Suggest two applications for a differential amplifier.

16 Explain why a tuned amplifier operates over a narrow frequency range.

17 What are the advantages of a tuned amplifier for use at radio frequencies?

18 If the differential mode gain of a differential amplifier is 50 and the common mode gain is −0.25, calculate the CMRR.

MULTIPLE CHOICE QUESTIONS

1 In a common-collector amplifier the output is taken from the:
A collector
B base
C positive supply
D emitter.

2 A transistor has an h_{fe} of 80. When the base current is increased by 15 µA, the collector current increases by:
A 15 mA
B 1.2 mA
C 1 mA
D 15 µA.

3 A common-emitter amplifier:
A has a voltage gain of 1
B has high output resistance
C has low input resistance
D is an inverting amplifier.

4 The voltage gain of a common-emitter amplifier with an emitter resistor depends on:
A the value of the emitter resistor only
B the value of the emitter resistance only
C the values of the emitter and collector resistors
D h_{fe}.

5 Compared with FET amplifiers, BJT amplifiers have:
A higher input resistance
B smaller voltage gain
C better linearity
D higher output resistance.

6 If the differential mode gain of a differential amplifier is 50 and the common mode gain is −0.25, the CMRR is:
A 50.25
B 200
C 23 dB
D 0.005.

7 The output of a tuned network shows a peak at the resonant frequency because:
A the capacitor and inductor have equal reactance
B the resonant network has high resistance
C the inductor has high Q
D resonance generates high voltages.

8 The transistors of a Darlington pair have current gains of 150 and 200. Their combined current gain is:
A 30 000
B 350
C 400
D 50.

9 The effects of stray capacitance are eliminated in a tuned amplifier because:
A capacitance has no effect at high frequency
B the capacitance acts as part of the resonant network
C the capacitance is cancelled by the inductance
D the transistor is connected so as to avoid capacitance effects.

10 The stabiliy of a common-emitter amplifier is improved if:
A we use a high-gain transistor.
B the base is biased with a resistor connected to the collector.
C the base is biased with resistors connected to the positive supply and the 0 V line.
D we use a high-power transistor.

11 A bypass capacitor connected in parallel with the emitter resistor of a common-emitter amplifier:
A improves stability.
B decreases gain.
C increases bandwidth.
D increases gain.

More on Transistors

The Companion Site has more questions on MOSFETs and BJTs. There are Calculator windows to help you find the answers.

JFET Amplifiers

Junction field effect transistors have been replaced by MOSFETs as the most commonly used field effect transistors. If you are working for an examination, check the subject specification to find out whether or not JFETs are included in it.

Like MOSFETs, JFETS are unipolar transistors, made in n-channel and p-channel versions (see Extension Box 15, p. 69). In this topic we describe an amplifier based on an n-channel JFET.

JFET AMPLIFIER

The JFET common-source amplifier shown below is very similar to the MOSFET amplifier described on p. 63 and operates in much the same way. The main difference is in the biasing of the gate.

FIGURE 10.1 A JFET common-source amplifier is biased by a single pull-down resistor.

In JFETs the gate is insulated from the substrate of the transistor by being reverse-biased. It is held negative of the source, causing a depletion region to form between the gate and the substrate, in the same way that it is formed in a reverse-biased diode. The usual way of biasing the gate is to connect a single high-value resistor (R1) between the gate and the 0 V line. This holds the gate very close to 0 V.

We then make the source a few volts *positive* of the gate by wiring a resistor (R3) between the source and the 0 V line. Relative to the source, the gate is negative.

The threshold voltage of the 2N3819 is -4 V, so a suitable quiescent gate voltage is in the region of -2 V. The gate can be biased at any voltage between the threshold and about $+0.5$ V. Above 0.5 V, the junction at the gate becomes forward-biased, the depletion region breaks down and the gate is no longer insulated from the body of the transistor.

Example

In the circuit (left) the quiescent current through the transistor is 0.5 mA. The voltage drop across R3 is:

$$v_{GS} = -i_D R_3 = -0.0005 \times 3900 = -1.95 \text{ V}$$

In practice, the voltage across R3 varies with the signal, making v_{GS} vary too. As the signal increases, v_{GS} increases, which tends to partly turn off the transistor and so reduces the signal. This is **negative feedback**, which reduces the gain of the amplifier. A large-value capacitor C3 is used to stabilise the voltage at the source of Q1, so reducing the feedback and increasing the gain.

Looking at this another way, the capacitor passes the signal through from the source to the 0 V line, leaving only the quiescent voltage at the source. For this reason, the capacitor is called a **by-pass capacitor**.

TRANSCONDUCTANCE

The transconductance of JFETs is generally much lower than that of MOSFETs. For the 2N3819, the transconductance $g_m = 0.4$ mS. This compares with 150 mS for the ZVN3306A MOSFET. We calculate the expected voltage gain as for the MOSFET common-source amplifier.

C   lower than −5 V

D   less than 0.5 V.

2   Compared with its MOSFET equivalent, the bandwidth of a JFET common-source amplifier is:

A   less

B   much wider

C   about the same

D   a little wider.

3   The output resistance of a common-source amplifier depends on:

A   the transconductance of the FET

B   the value of the output coupling capacitor

C   the value of the drain resistor

D   the gate-source voltage.

Operational Amplifiers

Operational amplifiers are precision, high-gain, differential amplifiers. They were originally designed to perform mathematical *operations* in computers, but nowadays this function has been taken over by digital microprocessors. Operational amplifiers (or 'op amps' as they are usually called) are still widely used in many other applications. An op amp can be built from individual transistors and resistors but practically all op amps are manufactured as integrated circuits. Dozens of different types of op amp are available with various combinations of characteristics.

TERMINALS

All op amps have at least five terminals:

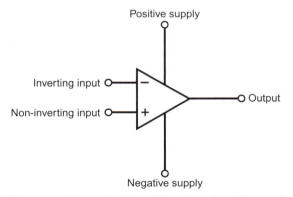

FIGURE 11.1 All operational amplifiers have these five terminals. Some may have more.

- **Positive and negative supply:** Most op amps run on a **dual supply** (or **split supply**). Typical supply voltages are ±9 V, ±15 V or ±18 V. Some op amps run on low voltages such as ±1 V or ±2 V. Certain op amps that are capable of accepting input voltages close to the supply rails can also run on a single supply, such as 2 V, up to 36 V. The power supply terminals are often omitted from circuit diagrams to simplify the layout.

- **Inverting and non-inverting inputs:** These are the inputs to the first stage of the amplifier, which is a differential amplifier. In this book we refer to these inputs as (+) inputs.
- **Output terminal.**

Amplifiers may also have two or three other terminals, including the **offset null** terminals.

PACKAGES

Most op amps are available as a single amplifier in an 8-pin integrated circuit package, and practically all have the standard pinout shown below.

FIGURE 11.2 Op amp integrated circuits have these pin connections. A circular dimple indicates which is pin 1.

Pins 1 and 5 are used for other functions, depending on the type. Many op amps are also available with 2 or 4 amplifiers in a single 14-pin integrated circuit package. They share power supply pins and usually do not have terminals for special functions.

DIFFERENTIAL OUTPUT

Assuming that the op amp is working on a dual supply (the mid-rail voltage is 0 V), output is positive when the (+) input voltage exceeds the (−) input voltage. It is negative when the (−) input voltage exceeds the (+) input voltage. It is zero when inputs are equal.

ACTIVITY — VOLTAGE COMPARATOR

Demonstrate the action of an op amp as a voltage comparator using the circuit below. The meters must be able to read both positive and negative voltages. Either use digital multimeters with automatic polarity, or centre-zero moving-coil meters.

FIGURE 11.6 Measuring the input offset voltage of an op amp.

Keep the (+) input constant (adjust VR1 to bring ME1 to +1 V, for example) and then adjust VR2 to set the (−) input at a series of values ranging from −6 V to +6 V.

Record the readings of the three meters in a table. Repeat with a different voltage on ME1, or hold the (−) input constant and vary the (+) input.

Write a summary interpreting your results. What is the reading on ME3 when the input voltages are equal? If ME3 does not show a reading of 0V, can you explain why? Does the output voltage ever swing

fully to +6 V or −6 V? How far does it swing in each direction? Repeat for another type of op amp.

INVERTING AMPLIFIER

Of the many applications of op amps, the two most commonly found are the inverting amplifier and the non-inverting amplifier. They have very different properties.

The inverting amplifier (Figure 11.7) has part of the output signal fed back to the inverting input. This is **negative feedback**, and the circuit has three interesting properties as a result.

EQUAL INPUT VOLTAGES

The (+) input has high input resistance. Only a minute current flows into it and so only a minute current flows through R_B. As a result, there is almost no voltage drop across R_B and the (+) input is practically at 0 V.

FIGURE 11.7 In an inverting amplifier circuit, the input goes to the (−) terminal and there is negative feedback to that terminal.

If v_{IN} is made positive, v_{OUT} becomes negative. There is a fall in voltage across the resistor chain R_A and R_F. At the v_{IN} end it is positive and at the v_{OUT} end it is negative. Somewhere along the chain the voltage is zero. As v_{OUT} falls, it pulls down the voltage at the (−) input until this is the point at which the voltage is zero.

If v_{OUT} falls any further, the (−) voltage becomes *less* than the (+) voltage and v_{OUT} begins to rise. If the voltage is exactly zero (taking this to be an ideal amplifier in which case we can ignore input offset voltage), v_{OUT} neither rises nor falls. Feedback holds it stable. The important thing is that the (−) input has been brought to the *same voltage* as the (+) input.

The same argument applies if v_{IN} is made negative. It also applies if the (+) input is held at any voltage above or below zero. This allows us to state a rule:

An inverting amplifier comes to a stable state in which the two inputs are at equal voltage.

GAIN SET BY RESISTORS

The diagram below shows typical resistor values. The voltages are shown for when v_{IN} is a constant 100 mV and the circuit has reached the stable state referred to above. Because of the high input impedance of the $(-)$ input, no current flows into it (ideally). All the current flowing along R_A flows on along R_F. Call this current i.

Because the $(-)$ input is at 0 V, the voltage drop across R_A is:

$$v_{IN} = iR_A$$

Similarly, the voltage drop across R_F is:

$$v_{OUT} = iR_F$$

Combining these two equations, we obtain:

Voltage gain, $A_V = \dfrac{v_{OUT}}{v_{IN}} = \dfrac{iR_F}{iR_A} = \dfrac{R_F}{R_A}$

FIGURE 11.8 The output voltage stabilises when the voltages at the $(+)$ and $(-)$ inputs are equal.

This leads to another rule for op amps:

The voltage gain of the circuit is set only by the values of the two resistors.

Example

An inverting op amp amplifier circuit has an input resistor R_A of 10 kΩ, and a feedback resistor R_F of 560 kΩ. Its voltage gain is:

$$A_V = \frac{R_F}{R_A} = \frac{560\ 000}{10\ 000} = 56$$

Although the op amp may have an open-loop gain of 200 000, the amplifier *circuit* has a gain of only 56. This is known as the **closed-loop gain**, and must always be substantially less than the open-loop gain of the op amp.

The open-loop gain is subject to tolerance errors resulting from differences arising during manufacture. But the closed-loop gain depends only on the precise values of the resistors. If 1% tolerance resistors are used the gain is precise to 1%.

Self Test

An inverting amplifier has $R_A = 2.2$ kΩ and $R_F = 820$ kΩ. Calculate its closed-loop gain.

The closed-loop gain obtained from a given combination of R_A and R_F can never be as great as the open-loop gain of the op amp.

VIRTUAL EARTH

If the $(+)$ input is at 0 V, the circuit stabilises with $(-)$ at 0 V too. The input current flows through R_A toward the $(-)$ input and an *equal* current flows on through R_F. Although no current flows into the $(-)$ input, the action is the same as if there is a direct path from the $(-)$ input to the 0 V rail. The rule is:

The $(-)$ input of an op amp connected as an inverting amplifier acts as a virtual earth.

Virtual earth is a helpful concept when analysing certain op amp circuits.

INPUT RESISTANCE

As a consequence of the virtual earth, a signal applied to the input has only to pass through R_A to reach a point at 0 V. Therefore, the input resistance of the amplifier in this example is only 10 kΩ. Although the input resistance *of the op amp* may be at least 2 MΩ and possibly as high as 10^{12} Ω, the input resistance *of the amplifier circuit* is nearly always much less.

Self Test

What is the input resistance of the amplifier described in the Self Test above?

If resistor values are being chosen for high gain, R_F is made large and R_A is made small. As a and R_A is omitted. The full v_{OUT} is fed back to the inverting input. As before, the circuit settles so that the inputs are at equal voltage. The non-inverting input is at v_{IN}, so the inverting input must be at v_{IN} too.

FIGURE 11.11   A voltage follower is useful as a buffer between circuits.

But the inverting input is at v_{OUT}, being connected directly to the output. As a result of this:

$$v_{OUT} = v_{IN}$$

Because the output voltage equals the input voltage, this is called a **unity-gain** amplifier. It is also known as a **voltage follower**. It has the same applications as the emitter follower and source follower transistor amplifiers, to act as a buffer between a circuit with high output resistance and a circuit with low input resistance.

The op amp is well suited as a buffer because the input resistance can be up to $1\,T\Omega$, and its output resistance is only $75\,\Omega$. The op amp follower is preferred to the transistor followers because its output follows input exactly, apart from a small error due to input offset voltage. In contrast, the transistor followers always have a much larger offset (0.7 V in the case of the emitter follower) and their gain is not *exactly* 1 (usually nearer to 0.9 in the source follower).

QUESTIONS ON OP AMPS

1 Describe the appearance of a typical operational amplifier packaged as an 8-pin integrated circuit.
2 How would you measure the open-loop gain of an op amp? What result would you expect to get?
3 Describe what is meant by slew rate. Why may we sometimes need to employ an op amp with high slew rate?

4 List the terminals present on all op amps and state their functions.
5 Describe how the open-loop gain of an op amp varies over the frequency range from 0 Hz (DC) to 100 MHz.
6 State two rules that are useful when designing op amp circuits.
7 Design a circuit for an op amp inverting amplifier with a closed-loop gain of -200, assuming an ideal op amp.
8 Design a circuit for an op amp non-inverting amplifier with a closed-loop gain of 101, assuming an ideal op amp.
9 Explain why op amp inverting amplifiers have a virtual earth.
10 What is input offset voltage, what is its effect, and how is it nulled?
11 Explain the meaning of (a) gain-bandwidth product, and (b) transition frequency.
12 Describe how an op amp voltage follower works and give an example of its applications.
13 Select an op amp from the table on p. 95, suitable for a battery-powered temperature alarm. The op amp compares a standard voltage with one produced by a temperature sensor and triggers the alarm when the selected temperature is reached.
14 An inverting amplifier based on a 355 op amp has $R_F = 100\,k\Omega$ and $R_A = 5.6\,k\Omega$. What is a suitable value of R_B? What is its input resistance? What is the open-loop gain of the 355? Estimate its closed-loop gain in this circuit at 1 kHz, 1 MHz, and 2.5 MHz.
15 Select an op amp from the table on p. 95 for use in a battery-powered domestic audio intercom system, explaining the reasons for your choice.
16 For what kind of application would you use an op amp as a non-inverting amplifier rather than as an inverting amplifier?
17 Describe the part played by the TL081 in the delay circuit on p. 41.

DATA — OPERATIONAL AMPLIFIERS

Op amps are listed by type number under the three kinds of input stage (see table on p. 95). BJT input op amps usually have the lowest input offset voltages. Bifet and MOSFET types have very high input resistance, and very low input bias current. Many types of MOSFET allow the output to swing to one or both supply rail voltages, but have lower slew rates than Bifet types. The op amps marked * in the table are also available in dual or quadruple versions. These

Type	Supply Voltage Range (V)	Open Loop Gain (dB)	Input Resistance (Ω)	Input Offset (mV)	Slew Rate (V/μs)	Full-Power Bandwidth (kHz)	Transition Frequency (MHz)	Output Swing (V)	Applications
BJT inputs									
741*	±5 to ±18	106	2 M	1	0.5	10	1	±13	General purpose
OP27	±4 to ±18	123	4 M	0.03	2.8	-	5	±11	Low input offset, low noise, precision
5539	±8 to ±12	52	100 k	2	600	48 000	1200	+2.3/−2.7	High slew rate and full-power bandwidth
Bifet inputs (JFET + BJT)									
LF355	±4 to ±18	106	1 T	3	5	60	2.5	±13	General purpose
TL071*	±2 to ±18	106	1 T	3	13	150	3	±13.5	Low supply voltage, low noise
TL081*	±2 to ±18	106	1 T	5	13	150	3	±13.5	Low supply voltage, general purpose
MOSFET inputs									
CA3130E†	±3 to ±8	110	1.5 T	8	10	120	4	O/P swings to both rails	wide bandwidth
CA3140E†	±2 to ±18	100	1 T	5	9	110	4.5	O/P swings to neg. rail	general purpose

versions save board space and simplify the pcb layout. Op amps marked † are suitable for running from a single supply voltage.

SELECTING OP AMPS

When selecting an op amp, input offset voltage, slew rate, and output swing are the usually the most important factors to consider. The open loop gain is nearly always far greater than the designer needs.

Sometimes high input resistance is important, but remember that the input resistance of many op amp circuits equals that of the input resistor. It is not dependent on the input resistances of the op amp IC itself.

Full-power bandwidth is important for radio-frequency circuits and high-speed digital signals. The transition frequency (p. 89) is a limitation that should be considered for signals of such high frequency.

Low noise op amps, such as the TL071, are often essential in applications where a sensor produces a signal of very low voltage, or when the signal consists of small variations in voltage. Normally the closed-loop gain is high and any noise in the signal is amplified along with the signal itself.

MULTIPLE CHOICE QUESTIONS

1 The input resistance of a bipolar op amp is typically:
   A 75 Ω
   B 75 kΩ
   C 200 kΩ
   D 2 MΩ.

2 The output resistance of an op amp is typically:
   A 75 Ω
   B 75 kΩ
   C 200 kΩ
   D 2 MΩ.

3 The open-loop gain of an op amp falls to 1 at:
   A 1 MHz
   B the transition frequency
   C 10 kHz
   D 75 kHz.

4 The input resistor of an inverting op amp is 12 kΩ and the feedback resistor is 180 kΩ. The input voltage is −0.6 V. The output voltage is:
   A 1.5 V
   B 9 V
   C −9.5 V
   D −16 V

5 In the amplifier described in Question 4, the value of the resistor between the (+) input and 0V should be:
   A 192 kΩ
   B 11.25 kΩ

voltage drop of v_2 along the chain and, because R3 and R4 are equal, the voltage at the (+) input is $v_2/2$.

FIGURE 12.2   A subtractor circuit produces an output equal to the difference between its inputs.

The voltage at the (−) input is $v_2/2$. For the same reasons as for the other amplifiers, the circuit comes to a stable state when the voltages at the (+) and (−) inputs are equal.

Consider the resistor chain R1/R2. This has a voltage v_1 at one end, v_{OUT} at the other end and $v_2/2$ in the centre between the two resistors. No current leaves the chain to flow into the (−) input. Because the resistors are equal in value there are equal voltage drops across them. The voltage drops are:

$$v_1 - v_2/2$$

and

$$v_2/2 - v_{OUT}$$

These are equal:

$$v_1 - v_2/2 = v_2/2 - v_{OUT}$$

Rearranging gives:

$$v_{OUT} = v_2 - v_1$$

The output voltage equals the difference between the input voltages.

The difference is amplified if R1 = R3 and R2 = R4, with R3 and R4 being larger than R1 and R3. Then:

$$v_{OUT} = (v_2 - v_1) \times \frac{R_2}{R_1}$$

Alternatively, the difference may be reduced in scale by making R_2 and R_4 smaller than R_1 and R_3.

INPUT RESISTANCE

A disadvantage of this circuit is that the input resistances of the two input terminals are unequal. When input voltages are equal, the (−) input of the amp is a virtual earth

so the input resistance at the v_1 input is R_1. But the input resistance at the v_2 input is R_3 and R_4 in series. If all resistors are equal, the input resistance of the v_2 terminal is double that of the v_1 terminal. If the resistances of R_2 and R_4 are large because the voltage difference is to be amplified, the input resistance of the v_2 terminal is much greater than that of the v_1 terminal.

The effect of v_1 having a smaller input resistance is to pull down all voltages applied to that terminal, making v_{OUT} smaller than it should be. In some applications, both input signals include certain amounts of mains hum or other interference picked up equally on the input cables (see next section). A differential amplifier has common-mode rejection (p. 77) so this normally does not matter. But, if the inputs have unequal input resistances, as in the difference amplifier circuit, the interference has a stronger effect on the input with the higher input resistance, and is not rejected.

MEDICAL INSTRUMENTATION

Difference amplifiers are often used for measuring the small electrical potentials generated within the human body. The activity of the heart is often explored by measuring the potentials generated by its beating muscle and displaying the results as an **electro-cardiograph**. Silver or platinum probes are attached to the wrists and leg. Difference amplifiers measure the potential differences between them.

Another common application is the measurement of potentials produced by the brain. The plotted results is an **electro-encephalogram**. High-gain circuits are needed because the brain is less active than the heart and produces potentials of only a hundred microvolts or so.

The difficulty in measuring potentials within the human body is not that they are small but that the potentials of interest are swamped by other potentials generated in the body and in the equipment.

The body is composed largely of conductive fluids. Electromagnetic fields such as those surrounding electrical equipment and wiring can induce electric currents within the fluids. Typically, these currents alternate at the mains frequency, 50 Hz. If you have ever touched a finger against an input terminal of an audio amplifier, you will probaby have heard a loud 50 Hz buzz. This is the kind of background noise from which we must extract the far weaker signals originating from the heart or brain.

Potentials coming from the heart or brain affect one of the inputs of the amplifier more than the other. They are differential-mode signals and are amplified. Potentials arising from electromagnetic interference affect both inputs equally. They are also equally likely to be picked up by the leads connecting the probes to the amplifier. They are common-mode signals, and are largely rejected, depending on the common mode rejection ratio (CMRR) of the amplifier.

Most op amps have a high CMRR. Of the types listed on p. 97, most have a CMRR in the region of 90−100 dB. This means that the differential-mode gain is 31 060 to 100 000 times greater than the common-mode gain. This allows the differential mode signal to be picked out clearly against a background of common-mode interference and other noise.

Some op amps are less suitable for this application as they have lower CMRRs. An example is the TL081 with a CMRR of 76 dB.

BRIDGE MEASUREMENTS

A number of measuring circuits are based on the Wheatstone Bridge. The bridge is made up of four resistances:

FIGURE 12.3 The basic Wheatsone bridge produces a voltage difference, v_{OUT}.

Think of the bridge as two potential dividers ACB and ADB, side by side. Currents i_1 and i_2 flow through the two dividers.

One or more of the resistances are adjustable, as will be explained later. Adjustments are made until the bridge is **balanced**. In this state, points C and D are at the same voltage, and $v_{OUT} = 0$ V.

Therefore:

$$v_{AC} = v_{AD} \quad \text{and} \quad v_{CB} = v_{DB}$$

and, by Ohm's Law:

$$i_1R_1 = i_2R_3 \quad \text{and} \quad i_1R_2 = i_2R_4$$

Dividing gives:

$$\frac{R_1}{R_2} = \frac{R_3}{R_4}$$

If the bridge is balanced and we know three of the resistances, we can calculate the fourth.

An example of the use of this technique is the measuring of mechanical strain by means of a **strain gauge**. The gauge is a thin metal foil etched to form a number of fine wires. This is embedded in a plastic film and cemented to the object (such as a steel girder) which is to be stressed.

FIGURE 12.4 When this strain gauge is stressed, the wires become longer and thinner, so their resistance increases.

There are various ways that one or more strain gauges may be included into a bridge. There are two identical gauges in the version below, one of which is stressed. The dummy is not stressed but is included in the bridge to compensate for changes of resistance of the gauge and its leads due to temperature.

FIGURE 12.5 In this version of the Wheatstone bridge, R2 consists of a fixed resistor and a variable resistor to allow the total resistance of R2 to be adjusted.

R1 and the fixed part of R2 consist of high-precision resistors. The variable part of R2 is a high-precision variable resistor with a calibrated knob.

Traditionally, points C and D are joined by a centre-reading microammeter. This displays a zero reading when the bridge is balanced. Instead, we can use an op amp connected as a difference amplifier. The variable resistor of R2 is adjusted until the output of the amplifier is zero. The high CMRR of the op amp makes it possible to detect small differences of potential between C and D, in spite of common-mode voltage signals picked up in the leads and other parts of the circuit.

To make a measurement, we subject the test piece (for example, the girder) to stress. We then adjust the variable resistor until the output of the amplifier is zero. We already know the resistance of R1 and the fixed portion of R2. We now read off the resistance of the variable portion of R2. The resistance of the dummy at the ambient temperature is taken from data tables. From these values we calculate the resistance of the strain gauge. Such equipment could be used for measuring forces, including weighing large loads.

The resistance of the strain gauge can then be read off in terms of mechanical strain by consulting data tables. Alternatively, it is possible to calibrate the equipment by subjecting the test piece to a range of known stresses and plotting a graph to relate strain to the setting of R2.

The difference amplifier is used for other measurements in which small variations of resistance of voltage must be determined against a background of common-mode interference. The platinum resistance thermometer depends on the increase with temperature in the resistance of a coil of platinum wire. These thermometers have high precision and can be used for temperatures ranging from $-100°C$ to over $1500°C$. However, the temperature coefficient of metals is very small, so a bridge method is used to obtain the required precision.

INTEGRATOR

The best way to understand what the circuit below does is to see how it behaves when a given input is applied to it. The (+) input is at 0 V, and the effect of negative feedback is to make the (−) input come to 0 V. It is a virtual earth, as described in Topic 9. If there is a constant input v_{IN}, current flows through R to the virtual earth. The current is $i = v_{IN}/R$, because the virtual earth causes the voltage across R to be equal to v_{IN}. In fact, this current does not flow into the (−) input but flows toward the capacitor.

FIGURE 12.6   In an integrator, the output falls as charge flows toward plate A of the capacitor.

Putting this into equations, we can say that if current i flows for t seconds, the charge Q flowing toward plate A is (using the relationship between Q, i, and t as in the definition of the coulomb):

$$Q = it$$

If this charge were to accumulate on Plate A, it would produce a voltage v on the plate which is (by definition of the farad):

$$v = Q/C$$

Combining these two equations:

$$v = it/C$$

Substituting the value of i:

$$v = v_{IN} \times t/RC$$

This assumes that the charge accumulates and that the voltage increases. But negative feedback holds the input at 0 V. To make this happen, the output v_{OUT} falls, pulling down the voltage at plate B, which in turn pulls down the voltage at plate A, and holding it at 0 V. The fall in output exactly compensates for the charge that accumulates on plate A:

$$v_{OUT} = -v$$

$$v_{OUT} = \frac{-1}{RC} \times v_{IN}t$$

If v_{IN} is constant, v_{OUT} equals v_{IN} multiplied by time t, multiplied by a scaling factor $-1/RC$. As time passes, v_{OUT} *decreases* at steady rate proportionate to $1/RC$. The larger the values of R and C, the more slowly it decreases.

Self Test

If v_{IN} is constant at 20 mV, R is 10 kΩ and C = 470 nF, what is v_{OUT} after 25 ms?

This discussion assumes that the capacitor is uncharged to begin with, that is, when t = 0. This is the purpose of the switch, which is closed to discharge the capacitor at the beginning of each run. The result of the *Self Test* problem (above) illustrates the fact that, with components of typical values, the output voltage rapidly falls as far as it can in the negative direction and the op amp then becomes saturated.

It also reminds us that quite a small input offset voltage can cause the circuit to swing negative in a second or so, even when v_{IN} is zero. Consequently it is important to use an op amp with small input offset voltage, and to use the offset null adjustment, as shown on p. 87.

If the input voltage is varying, we can work out what will happen by thinking of the time as being divided into many very short time intervals of length δt. We can consider that the voltage is constant during each of these, but changes to a new value at the start of the next interval.

During each interval, charge accumulates according to the constant-voltage equation. The total charge, and also the fall in output voltage at any given time, is the sum of the charges and voltage drops that have occurred during all previous intervals. For each interval, the voltage drop is $v_{IN}\delta t$ multiplied by $-1/RC$. For all the intervals between zero time and time t, we obtain the total voltage drop by calculating:

$$v_{OUT} = \frac{-1}{RC} \int_0^t v_{IN} dt$$

This is the *integral* of v_{IN} with respect to time, from time zero to time t, multiplied by the factor $-1/RC$.

If we divide v_{OUT} by the factor $-1/RC$ we obtain the value of the integral, and this is the average value of v_{IN} during the period from 0 to t. In this way, the circuit is a useful one for averaging a voltage over a period of time.

RAMP GENERATOR

The circuit has another important application, as a **ramp generator**. If we apply a constant positive or negative voltage to an integrator, v_{OUT} ramps down or up at a steady rate. Ramping voltages are often used in timing circuits. The output voltage is proportional to the time elapsed since the beginning of the ramp.

INVERTING SCHMITT TRIGGER

This circuit in Figure 12.7 has the same kind of action as the trigger circuit on p. 18. Note that the op amp uses a single supply in this application, so the type chosen must be suitable for use in this way. Preferably, the output should be able to swing to equal either supply voltage, which is why a CA3130E has been selected.

R1 and R2 form a potential divider and, because they are equal in value, the voltage at point A is 6 V, half-way between the two supply rails. Low-value resistors are selected for R1 and R2 so that the voltage at A is not seriously affected by current flowing into or out of the divider.

A second potential divider is formed by R3 and R4. This has one end connected to the op amp output and the other end at point A. These resistors have relatively high values so that they do not draw excessive current from the R1-R2 divider. The output of this

FIGURE 12.7 An inverting Schmitt trigger uses positive feedback to give it its 'snap' action.

second divider goes to the (+) input of the op amp. Since the op amp input resistance is 1 TΩ, this draws virtually no current from the second divider and the voltage at the (+) terminal will be as calculated.

SCHMITT ACTION

The graph in Figure 12.8 (taken from a simulation) displays the action of the circuit when the input is a sawtooth waveform, amplitude 4 V, centred on +6 V.

Starting with v_{IN} (light grey) at its lowest peak (2 V), the op amp is saturated, and its output (black) is high. Feedback through R3 pulls the (+) input voltage (dark grey) above the halfway value to 8 V. This is the **upper threshold voltage** (UTV). Consequently, v_{IN} has to rise to 8 V before the (−) input exceeds the (+) input and the output of the amplifier swings low.

FIGURE 12.8 As the input (light grey) to the trigger circuit ramps slowly up and down, the output (black) alters abruptly. Changes in the threshold are plotted in dark grey.

As soon as v_{IN} exceeds 8 V, the voltage at the (+) input is greater than that at the (−) input. The output begins to swing low, toward 0 V. As it swings, it pulls the voltage at the (−) input down below 8 V, so increasing the differential between the two inputs. This accelerates the downward swing and produces the characteristic 'snap' action of the Schmitt trigger. Note the inverting action.

The other result is that, once the swing has begun, any slight decrease in v_{IN} can have no effect. Once begun, the change in state can not easily be reversed.

Once the output has fallen to 0 V, the potential divider R3-R4 has 0 V at one end and the halfway voltage (6 V) at the other. The voltage at the (+) input is now below the half-way value, at 4 V. This is the **lower threshold voltage** (LTV). Consequently, as v_{IN} falls from its positive peak, it has to fall below 4 V before the trigger changes back to its original state. Then v_{IN} is less than the LTV and the output goes high.

With a sawtooth input (or any other periodic waveform such as a sinusoid), the action continues as above, the output falling every time input exceeds the UTV and rising every time input drops below the LTV. The difference between UTV and LTV is the **hysteresis** of the trigger, 4 V in this example.

Schmitt triggers are often used for squaring irregular waveforms. Square waves, pulses and digital signals become distorted in shape after transmission over a long distance. They may also pick up interference from electromagnetic and other sources. Passing the signal through a Schmitt trigger removes irregularities and restores the square shape.

CALCULATING THRESHOLDS

As an example, we will calculate the thresholds of the trigger circuit just described. The first step is to calculate the voltage at point A. This is easy in this example because R1 = R2. The voltage at A, which we will call V_M, is half-way between 0 V and 12 V, so $V_M = 6$ V. If R1 and R2 are unequal, and the supply voltage is V_S, the formula is:

$$V_M = V_S \times \frac{R_2}{R_1 + R_2}$$

This is derived from the usual potential divider formula. Looking at the second potential divider (R3-R4) we see that, when v_{OUT} is high, the end connected to A is at V_M, while the other end is at V_S. The voltage drop across the divider is $V_S - V_M$. This is divided into two parts depending on the ratio between R_3 and R_4. The upper threshold voltage is the voltage V_U at the point between R3 and R4.

Figure 12.9 shows this potential divider and the voltages concerned. In this state v_{IN} has fallen below LTV, and v_{OUT} has swung up to become equal to the positive supply V_S. Now v_{IN} has to rise above the upper threshold V_U before the trigger will change state again.

FIGURE 12.9 The potential divider when v_{IN} is below the lower voltage threshold.

From Figure 12.9, V_U equals V_M plus the voltage rise across R4. Using the potential divider formula:

$$V_U = V_M + (V_S - V_M) \times \frac{R_4}{R_3 + R_4}$$

In the example the formula can be simplified, because $V_M = V_S/2$:

$$V_U = V_M \left(1 + \frac{R_4}{R_3 + R_4}\right)$$

Substituting the values used in the example, and working in kilohms.

$$V_U = 6(1 + 200/600) = 6 \times 1.333 = 8 \text{ V}$$

The UTV is 8 V, as shown by the graph.

When v_{IN} rises above the UTV, v_{OUT} swings down to the negative supply, which is 0 V in this single-supply circuit.

FIGURE 12.10 The potential divider when v_{IN} is above the upper voltage threshold.

The LTV is V_M, minus the voltage across R4:

$$V_L = V_M \left(1 - \frac{R_4}{R_3 - R_4}\right)$$

In the example, and working in kilohms:

$$V_L = 6(1 - 200/600) = 6 \times 0.667 = 4 \text{ V}$$

The LTV is 4 V, as shown by the graph.

The same formulae apply when V_M is *not* half of the supply. For example, suppose that we reduce R1 to 1.3 kΩ, which raises V_M to approximately 7.5 V. The thresholds become:

$$V_U = 7.5 + (12 - 7.5) \times 200/600 = 9 \text{ V}$$
$$V_L = 7.5(1 - 200/600) = 5 \text{ V}$$

Note that the thresholds are no longer symmetrical about V_M, but the hysteresis remains unchanged at 4 V.

DESIGNING AN INVERTING TRIGGER

It can be seen from the discussion above that the size of the hysteresis is decided by the values of R3 and R4. The location of the hysteresis band is decided by R1 and R2, but it is not necessarily symmetrical about V_M.

The calculation assumes that we can ignore the effects of current drawn from the R1-R2 divider to supply the R3-R4 divider. It also assumes that the output swings fully to the positive and negative supply rails. Many op amps do not allow the output to swing as far as the rails (see table, p. 95), and the extent of swing may vary in different op amps. If such op amps are used, it is preferable to measure the maximum swings before calculating the final values of resistors.

The design process begins with the selection of the threshold values V_U and V_L. The other known value is the supply voltage V_S (alternatively, the limits to which v_{OUT} swings). In order to calculate resistor values we first need to calculate V_M.

Referring to the voltage diagram on p. 105, and representing $R_4/(R_3 + R_4)$ by R to simplify the equations:

$$V_U = V_M + R(V_S - V_M) \qquad [Equation\ 1]$$

Referring to the voltage diagram on this page:

$$V_L = V_M - RV_M \qquad [Equation\ 2]$$

Subtracting Equation 2 from Equation 1:

$$V_U - V_L = RV_S$$

From this we obtain:

$$R = \frac{V_U - V_L}{V_S}$$

This gives us the data we need for calculating V_M. Equation 2 above can be re-written:

$$V_L = V_M(1 - R)$$

From which we obtain:

$$V_M = \frac{V_L}{1 - R}$$

Based on the equations above, the routine for designing an inverting trigger is as follows:

1 Decide on values required for V_U and V_L.
2 Calculate R, using:

$$R = \frac{V_U - V_L}{V_S}$$

3 Calculate V_M, using:

$$V_M = \frac{V_L}{1 - R}$$

4 Find two resistors R1 and R2, which total a maximum of 5 kΩ, to give the required value of V_M. This is most easily done by estimating a value for R2, then calculating R1:

$$R_1 = R_2\left(\frac{V_S}{V_M} - 1\right)$$

5 Find two resistors R3 and R4, which total a minimum of 50 kΩ, to give the required value of R. This is most easily done by estimating a value for R4, then calculating R3:

$$R_3 = \frac{R_4(1 - R)}{R}$$

Example

Design an inverting Schmitt trigger, using the circuit in Fig. 12.7. The supply voltage is 9 V, LTV = 3 V, UTV = 4 V.

We follow steps 2 to 5 above:

2 $R = (4 - 3)/9 = 1/9$
3 $V_M = 3/(8/9) = 3.375$ V
4 Let $R_2 = 2.2$ kΩ, then, working in kilohms:
  $R_1 = 2.2(9/3.375 - 1) = 3.667$ kΩ.
5 Let $R_4 = 200$ kΩ, then, working in kilohms:
  $R_3 = 200 (1 - 1/9)/(1/9) = 1600$ kΩ = 1.6 MΩ.

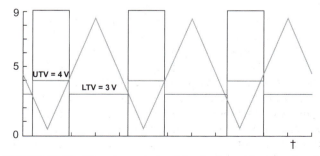

FIGURE 12.11   The values calculated for the four resistors produce a trigger action with UTV = 4 V and LTV = 3 V.

As a check on the calculations, the graph in Figure 12.11 was plotted for a simulated trigger circuit based on these results.

The method above can be used for any op amp that swings its output fully to the supply rails. It yields approximate results for op amps which do not swing to the rails, provided that the operating voltage is near the maximum for the op amp used.

Self Test

Design an inverting trigger for an op amp running on a 15 V single supply, so that UTV = 10 V and LTV = 7.5 V.

NON-INVERTING SCHMITT TRIGGER

This has a circuit similar to the inverting trigger, except that the connections to the op amp inputs are reversed. The input signal v_{IN} goes to the non-inverting (+) input, by way of the R3-R4 potential divider.

The point A, at V_M, is connected to the inverting input. The effect of these changes is to make v_{OUT} swing high as v_{IN} rises above the upper threshold. It swings low as v_{IN} falls below the lower threshold.

FIGURE 12.12 A non-inverting Schmitt trigger also uses positive feedback.

In this circuit the R1-R2 potential divider is connected only to the (−) input of the op amp. This draws a very small current, so large-values resistors can be used for R1 and R2. In the circuit above we have used 220 kΩ resistors instead of the 2.2 kΩ resistors used in the inverting trigger.

It is obvious from the graph that the action of this circuit differs in several ways from that of the inverting trigger (see graph, Fig. 12.13). The graph shows

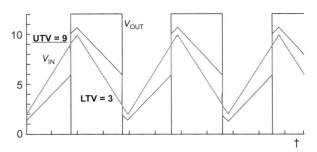

FIGURE 12.13 The (+) input to the op amp (dark grey line) is a sawtooth like v_{IN} (light grey) but is shifted up or down by 4.2 V every time it reaches 6 V. This happens because of v_{OUT} pulling up or down every time the op amp changes state.

the action of the circuit when the input is a sawtooth waveform, amplitude 4 V, centred on +6 V.

Starting with v_{IN} (light grey) at its lowest peak, the op amp is saturated and its output (black) is low.

The R3-R4 divider in Fig 12.14 has one end at v_{IN} and the other end at 0 V. As v_{IN} rises, the voltage at the (+) input of the op amp (dark grey, call this v_+) rises too. It is lower than v_{IN} because v_{OUT} is low and pulls it down. As v_{IN} rises, v_+ stays below it, until v_+ reaches v_- the voltage of the R1-R2 divider. The level of v_{IN} at this stage is the upper threshold voltage (UTV).

As v_{IN} exceeds UTV, v_{OUT} swings sharply to the positive supply (V_S). Instead of being pulled down, v_+ is now pulled up, clearly above UTV. Small reversals in the rise of v_{IN} can not affect the circuit after this. It is triggered.

When v_{IN} falls, v_+ falls too but stays a little higher than v_{IN}. This is because v_{OUT} is pulling it up. Eventually, v_{IN} falls to the lower threshold level (LVT). At this stage, v_+ falls below v_-. The circuit is triggered back to its original state, with v_{OUT} low.

THRESHOLDS

As before, it is helpful to sketch the R3-R4 divider and mark the voltages on it. As before, we simplify

FIGURE 12.14 The potential divider when v_{IN} is below the upper threshold voltage.

the equations by putting $R = R_4/(R_3+R_4)$. The result is shown below and we can see that:

$$v_+ = v_{IN} - v_{IN} \times R$$

Collecting terms gives:

$$v_+ = v_{IN}(1 - R)$$

v_{IN} rises until v_+ equals v_-, the voltage at the $(-)$ input of the op amp. At that point:

$$v_{IN} = \frac{v_-}{1 - R}$$

This is the upper threshold voltage, at which v_{OUT} swings from low to high.

The drawing below shows the divider after v_{IN} has passed through the UTV.

FIGURE 12.15 The potential divider when v_{IN} is above the lower threshold voltage.

Adding voltage drops:

$$v_+ = v_{IN} + (V_S - v_{IN})R$$

From which we deduce that, at the LTV:

$$v_{IN} = \frac{v_- - V_S R}{1 - R}$$

In the example, first calculate R, working in hundreds of kilohms:

$$R = 2/6 = 0.333$$

At the UTV, given that $V_S = 12$ V, $v_- = 6$ V, and working in hundreds of kilohms:

$$v_{IN} = \frac{6}{1 - 0.333} = 9 \text{ V}$$

This agrees with the value obtained in the simulation. At the LTV:

$$v_{IN} = \frac{6 - 12 \times 0.333}{1 - 0.333} = 3 \text{ V}$$

The thresholds are 9 V and 3 V, as shown in the diagram, and the hysteresis is 6 V.

DESIGNING A NON-INVERTING TRIGGER

The hysteresis is decided by the values of R3 and R4. The location of the hysteresis band is determined by R1 and R2, but it is not necessarily symmetrical about v_-.

The design process begins with the selection of the threshold values V_U and V_L. The supply voltage V_S is known. The quantity R is defined as on p. 103.

Referring to the voltage diagram in Figure 12.15:

$$V_U = v_+ + Rv_U \qquad [Equation\ 1]$$

Referring to the diagram on this page:

$$V_L = v_+ - R(V_S - v_L) \qquad [Equation\ 2]$$

Subtracting terms in Equation 1 from corresponding terms in Equation 2:

$$V_U - V_L = R(V_U - V_L + V_S)$$

From this we obtain:

$$R = \frac{V_U - V_L}{V_U - V_L + V_S}$$

This gives us the data we need for calculating v_+ (and hence v_-). From Equation 1:

$$v_+ = V_U(1 - R)$$

Based on the equations above, the routine for designing a non-inverting trigger is as follows:

1 Decide on values required for V_U and V_L.
2 Calculate R, using:

$$R = \frac{V_U - V_L}{V_U - V_L + V_S}$$

3 Calculate v_+ using:

$$v+ = V_U(1 - R)$$

This is also the value of v_-.
4 Find two resistors R1 and R2, to give the required value of v_-. This is most easily done by estimating a value for R2, then calculating R1:

$$R_1 = R_2\left(\frac{V_S - v_+}{v_+}\right)$$

5 Find two resistors R3 and R4 to give the required value of R. This is most easily done by estimating a value for R4, then calculating R3:

$$R_3 = R_4\left(\frac{1 - R}{R}\right)$$

Example

Design a non-inverting Schmitt trigger operating on a 15 V supply, with LVT = 6 V and UVT = 11 V.
Taking the calculation from step 2, onward:

2 $R = (11-6)/(11-6+15) = 0.25$

3 $v_+ = 11(1-0.25) = 8.25$

4 Let $R_2 = 200$ kΩ, then, working in kilohms:

$$R_1 = 200(15 - 8.25)/8.25 = 164 \text{ k}\Omega$$

5 Let $R_4 = 200$ kΩ, then, working in kilohms:

$$R_2 = 200(1 - 0.25)/0.25 = 600 \text{ k}\Omega$$

Self Test

Design a non-inverting trigger for an op amp running on a 6 V single supply, so that UTV = 4.5 V and LTV = 2 V.

As a check on the calculations, the graph below was plotted for a simulated trigger circuit based on this example.

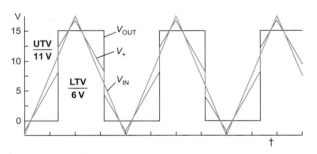

FIGURE 12.16 The input signal is a sawtooth, amplitude 10 V, centred on +7.5 V. The output is a square wave, amplitude 7.5 V, centered on +7.5 V.

RAMP AND SQUARE-WAVE GENERATOR

This is an application of the inverting Schmitt trigger. The block diagram in Figure 12.17 shows its main features.

The circuit repeatedly charges and discharges a capacitor through a resistor. When v_{OUT} is high, current flows through R to charge C. The voltage across C, and hence the input v_{IN} to the trigger, rises until it reaches the UTV of the trigger. Then the output swings sharply low, to 0 V.

FIGURE 12.17 A resistor-capacitor network connected to an inverting Schmitt trigger forms a simple square-wave generator.

With v_{OUT} at 0 V, the capacitor discharges through R. Discharge continues until the voltage across C, and the voltage at v_{IN} has fallen to the LTV. At this point v_{OUT} swings high again.

The result of this action is that v_{OUT} swings repeatedly between 0 V and the supply voltage. It produces a continuous square wave. The graph below shows the output from such a circuit, based on the trigger circuit Fig 12.7. The additional components are the resistor R, value 1 MΩ and the capacitor C, value 1 μF.

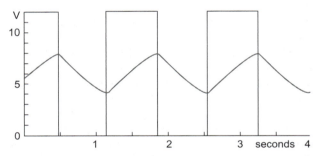

FIGURE 12.18 The curve for v_{OUT} is a square wave. The curve for v_{IN} is close to a sawtooth, but the segments are slightly curved.

The diagram below shows the circuit in more detail:

FIGURE 12.19 This trigger circuit is exactly the same as that in Fig 12.7 with the addition of the timing components R and C.

RAMP GENERATOR

Instead of taking the output of this circuit from the op amp output (v_{OUT}), we can take it from the point marked v_{IN}. This is plotted in grey on the graph opposite. The voltage ramps up and down, producing a sawtooth wave. However, the segments of the waveform are really segments of an exponential curve. Provided that the difference between UVT and LVT is not too large, the segments are close enough to straight lines for most applications.

Taking the output from the point labelled v_{IN} may draw an appreciable current from this circuit. This will make the charging times too long and the discharging times too short. In practice, the input terminal of a unity gain voltage follower (p. 93) is connected to this point. This provides an ample source of current for feeding the sawtooth signal to another circuit.

TIMING

Charging and discharging a capacitor takes time. The greater the capacitance of C and the larger the resistance of R, the longer it takes to charge and discharge the capacitor.

The frequency of the waveforms thus depends on the values of R and C. It also depends on the hysteresis of the trigger. The greater the hysteresis, the larger the rise and fall of v_{IN} and the longer it takes.

The times taken to charge to or discharge from certain key points on the exponential curves depends solely on the time constant, t, which equals RC. Timing is independent of the supply voltage. It follows that, given the threshold levels set by resistors R1 to R4, the frequency of the output signals depends only on the values of R and C. This is a useful feature of this circuit, for its frequency does not vary with changes in supply voltage.

Measurements on the plot opposite show that the period of the signals is almost exactly 1.4 s. The time constant of the circuit is:

$$t = RC = 1 \times .1 = 1 \text{ s} \quad [\text{M}\Omega \times \text{F}]$$

From this we deduce that the period is 1.4 time constants. Testing the circuit with other values of R and C shows that this is holds true. However, period depends also on the thresholds of the trigger, a topic which we study in the next section.

THRESHOLDS AND PERIOD

The thresholds for trigger circuits with the resistor values shown in the diagram (p. 93) are:

$$\text{UTV} = 8 \text{ V}$$
$$\text{LTV} = 4 \text{ V}$$

These values apply when $V_S = 12$ V. Putting the thresholds as fractions of the supply voltage we have:

$$\text{UTV} = 0.667$$
$$\text{LTV} = 0.333$$

During the charging phase we need to calculate the time required to charge C from $0.333V_S$ to $0.667V_S$.

The Extension Box 6 on pp. 37−38 shows how to calculate times when charging from 0 V up to to a chosen fraction of the supply voltage. Here we are charging from one fraction to another.

During charging, t_p is the time taken to charge from 0 V to pV_S, p having any value between 0 and 1.

Therefore, from the equation on p. 37:

$$1 - e^{\frac{-t_p}{RC}} = p$$

and therefore:

$$\frac{-t_p}{RC} = \ln(1 - p)$$

$$-t_p = \ln(1 - p)RC$$

During charging, the time to charge from 0 V to the LTV, where p = 0.333 is:

$$t_{0.333} = \ln(1 - 0.333)RC = 0.405RC$$

The time taken to charge from 0 V to the UTV, where p = 0.667 is:

$$t_{0.667} = \ln(1 - 0.667)RC = 1.100RC$$

The time taken to charge from LTV to UTV is the difference between these two times, so:

$$\text{Charging time} = (1.100 - 0.405)RC = 0.695RC$$

Now to consider discharging, for which the equation is:

$$-t_p = \ln(p)RC$$

The time to discharge from V_S to the UTV, where p = 0.667 is:

$$t_{0.667} = -\ln(0.667)RC = 0.405RC$$

The time taken to charge from V_S to the LTV, where p = 0.333 is:

$$t_{0.333} = -\ln(0.333)RC = 1.100RC$$

The time taken to discharge from UTV to LTV is the difference between these two times, so:

Discharging time $= (1.100 - 0.405)RC = 0.695RC$

Charging and discharging times are equal because, in this trigger, UTV and LTV are symmetrical about $V_S/2$. This does not necessarily apply to other trigger circuits.

The total length of one cycle is equal to the charging time plus the discharging time:

$$\text{Period} = 2 \times 0.695RC = 1.39RC$$

The period is 1.39 time constants. This confirms the result obtained by measuring the graph taken from the simulation (p. 106).

QUESTIONS ON OP AMP APPLICATIONS

1 The output of a sensor has high output resistance. Design a circuit (which may have more than one op amp in it) to take the output from the sensor, subtract 1.5 V from it and display the result on a multi-meter.

2 Design a circuit to mix three audio signals, two at equal amplitude and the third signal at half its original amplitude.

3 Explain why an op amp difference amplifier is used for measuring a small voltage signal against a background of interference. Illustrate your answer by quoting two practical examples.

4 Describe a technique for using a Wheatstone bridge to measure an unknown resistance.

5 Explain how a bridge may be used to measure temperature, using a platinum resistance thermometer as sensor.

6 Explain how an op amp integrator works when the input voltage is constant.

7 How can an op amp integrator be used as a ramp generator?

8 An op amp integrator circuit has R = 22 kΩ and C = 560 nF. Its output rises from 0 V to 125 mV in 40 s. What is the average input during this period?

9 Explain how an op amp integrator is used to measure the average value of the input voltage over a period of time.

10 Explain the action of an op amp inverting Schmitt trigger. Describe two examples of applications of this type of trigger circuit.

11 Explain the working of an op amp non-inverting Schmitt trigger circuit.

12 Calculate the upper and lower threshold voltages of this Schmitt trigger circuit:
How much is the hysteresis of this trigger (Fig 12.20)?

FIGURE 12.20

13 Design an op amp inverting Schmitt trigger that has its upper threshold voltage at 10 V and its lower threshold at 4 V. The supply voltage is 15 V.

14 Calculate the upper and lower threshold voltages of the Schmitt trigger circuit below:

FIGURE 12.21

In what way does the action of this trigger differ from that in Q. 12?

15 Design an op amp non-inverting Schmitt trigger that has its upper threshold voltage at 7.5 V and its lower threshold at 3.5 V. The supply voltage is 9 V.

16 Calculate the upper and lower threshold voltages of this trigger circuit:

FIGURE 12.22

17 Calculate the time constant of the circuit of Q. 16. How long is its period, in time constants? What is its period in seconds and its frequency in hertz?

18 Draw a circuit diagram to show how to use this circuit as a ramp generator.

MULTIPLE CHOICE QUESTIONS

1 An op amp used in an integrator circuit should have:
   A low input offset voltage
   B high open-loop gain
   C low slew rate
   D high input resistance.

2 An op amp integrator receives a sinusoidal input, amplitude 50 mV. Assuming the integrator does not have time for its output to swing fully in either direction, its output after three cycles of the sinusoid is:
   A 50 mV
   B 150 mV

C 0 V
D −25 mV.

3 An inverting Schmitt trigger circuit changes state when:
   A a rising input goes just above the upper threshold
   B a rising input goes just below the lower threshold
   C a falling input goes just below the upper threshold
   D the upper and lower thresholds are equal.

4 An op amp used in a Schmitt trigger should have:
   A low input offset voltage
   B high input resistance
   C low output resistance
   D high common-mode rejection ratio.

5 The difference between the upper and lower thresholds of a Schmitt trigger is known as:
   A feedback
   B input offset voltage
   C input bias current
   D hysteresis.

Active Filters

Active filters are another important application of operational amplifiers.

When a signal of mixed frequencies is passed through a filter circuit, some of the frequencies are reduced in amplitude. Some may be removed altogether. The simplest filter consists of a resistor and a capacitor (p. 44). As it is built from two passive components it is known as a **passive filter**.

> **Passive Component**
>
> One that does not use an external power source. Examples are resistors, capacitors, inductors, and diodes.

Other passive filters may consist of a network of several resistors and capacitors, or may be combinations of resistors and inductors, or capacitors and inductors.

A passive filter must have at least one reactive component (a capacitor or an inductor) in order to make its action frequency-dependent.

Passive filters are often used, but a large number of individual filters must be joined together to effectively cut off unwanted frequencies. Such filters take up a lot of space on the circuit board and can be expensive to build. When filtering low frequencies the inductors are necessarily large, heavy and expensive.

Another serious disadvantage of passive filters is that passive components can only *reduce* the amplitude of a signal. A signal that has been passed through a multi-stage passive filter is usually very weak. For these reasons, we prefer **active filters**, based on active components, especially op amps. When these are used we can obtain much sharper cutoff with relatively few filtering stages, there is no need to use inductors, and the filtered signal may even show amplitude gain.

> **Active Component**
>
> One that uses an external power source. Examples include transistors, and operational amplifiers.

FIRST-ORDER ACTIVE FILTER

The diagram below shows one of the simplest active filters. It consists of a resistor-capacitor passive filter followed by a non-inverting amplifier to restore or possibly increase the amplitude of the signal.

FIGURE 13.1   In a first-order active filter, the signal from the RC filter network is amplified by a non-inverting op amp amplifier.

With the values shown, the voltage gain of the amplifier is 1.56. The graph below displays the signals in this filter when it is supplied with a 1 V sinusoid at 1 kHz.

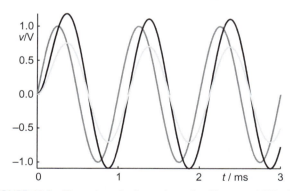

FIGURE 13.2   The action of a first-order active filter on a 1 kHz sinusoid (dark grey). The RC filter network reduces its amplitude and delays its phase (light grey). The amplifier then increases the amplitude (black).

The RC network filters the input sinusoid (dark grey line). At the output of the network (light grey) the filtered signal has an amplitude of only 0.7 V. This is not the only change. The output reaches its peaks 125 μs behind the input. This is a delay of 1/8 of the period of the signal. In terms of angle, this is 360/8 = 45°. Because it is a delay, it has negative value. The output is *out of phase* with, or *lags* the input, by −45°. The black curve represents the output of the op amp.

With the values shown for R_F and R_A, the gain of the amplifier is 1.56. The output from the RC network has amplitude 0.7 V, so the overall gain of the active filter is 0.7 × 1.56 = 1.1. The amplitude of the output is 1.1 V.

Gain

Gain of non-inverting amplifier = $(R_A + R_F)/R_A$

The amplifier does not change the phase any further, so the output lags the input by 45°.

FREQUENCY RESPONSE

The frequency response (see Topic 5) is plotted below.

The plot (black) shows that amplitude is the full 1.56 V up to about 300 Hz. From then on amplitude begins to fall off slightly, reaching 0.7 V (equivalent to −3 dB, or half power) at 1 kHz. The bandwidth of the filter is 1 kHz.

The frequency of f_C, the −3 dB point or cut-off point, depends on the values of the resistor and capacitor.

With the values of R and C shown in in Fig 13.1, the cut-off point is:

$$f_C = \frac{1}{2\pi RC}$$

$$f_C = 1/(2\pi \times 1.6 \times 100) = 1 \text{ kHz} \qquad [1/(k\Omega \times nF)]$$

As frequency is increased beyond 1 kHz, the amplitude rolls off even faster and eventually the curve plunges downward as a straight line. If we read off frequency and amplitude at two points on this line we find that for every doubling of frequency, the amplitude falls by 6 dB. We say that the roll-off is −6 dB per octave.

Octave

An interval over which frequency doubles or halves.

Another way of expressing this is to measure how much the amplitude falls for a ten-times increase in frequency, or decade. In the plot, the roll-off is 20 dB per decade.

Note that these roll-off rates are characteristic of single-stage passive or active filters and do not depend on the values of the capacitor and resistor.

The plot also shows how the phase change of the output varies with frequency. The scale shows phase angle on the right margin of the figure. We can see that the phase change at 1 kHz is −45°, confirming what we found in the output plot in Fig 13.3.

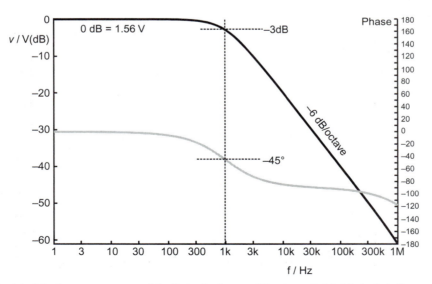

FIGURE 13.3 The Bode plot of the frequency response of the first-order lowpass filter shows that it falls to its half-power level (amplitude 0.7 V) and has phase lag of 45° at the cut-off.

Self Test

A first-order lowpass filter has a 47 kΩ resistor and a 220 pF capacitor. What is its cut-off frequency?

A phase lag of −45° at the cut-off point is another characteristic of single-stage passive or active filters. The phase lag is less at lower frequencies. At 0 Hz (DC) the phase change is zero. It increases at frequencies above the cut-off point.

HIGHPASS FIRST-ORDER ACTIVE FILTER

The lowpass filter is converted to a highpass filter by transposing the resistor and capacitor.

FIGURE 13.4 A first order highpass active filter consists of a high-pass passive filter followed by a non-inverting amplifier.

When this is done, the frequency response has the appearance shown below:

The amplitude curve rises up from 0 Hz at +6 dB per octave, or +20 dB per decade. The equation for calculating the cut-off point is the same for both low-pass and highpass filters, so this is 1 kHz as before. Beyond the cut-off point, amplitude rises to 0 dB above about 4 kHz.

The phase response curve shows that the filter has a 45° *lead* at its cut-off frequency. This is the opposite of the phase change found in the lowpass filter.

Looking more closely at the frequency response, we note that the amplitude starts to fall again from about 400 kHz upward. This is not the effect of the RC filter but is due to the fall in gain of the op amp at high frequencies. The transition frequency of the TL081C is 3 MHz. If we were to plot the response at higher frequencies, we should see a steep roll-off.

In practice, if we extend the operation of this filter into higher frequencies it acts as a bandpass filter.

DESIGNING FIRST-ORDER FILTERS

1 Decide on a capacitor value.
2 Apply the equation: $R = 1/2\pi f_C C$.

Example

To design a lowpass filter with cut-off point at 40 kHz.
1 For stability, try the calculation for a 220 pF capacitor, so that a ceramic plate capacitor with zero tempco can be used.
2 Applying the equation, $R = 18086\ \Omega$.
Use an 18 kΩ, 1% tolerance, metal film resistor. For a highpass filter, exchange the resistor and capacitor.

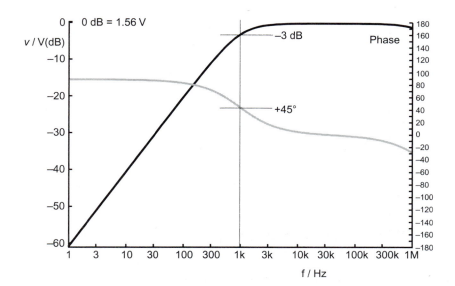

FIGURE 13.5

SECOND-ORDER ACTIVE FILTER

Adding an extra stage to the first-order lowpass filter allows feedback to be introduced. The circuit below has two RC filters, one of which is connected to 0 V, as in the first-order lowpass filter. The other is part of a feedback loop from the filter output.

FIGURE 13.6   The second-order active filter has positive feedback from the output to the capacitor of the first stage.

The feedback action makes the filter resonate at frequencies around the cut-off frequency. Resonance is not strong, but there is enough extra feedback around this frequency to push up the 'knee' of the frequency response curve.

FIGURE 13.7   The Bode plot of the response of the second-order filter (black) also shows the amplitude of the feedback signal (grey), which pushes up output at the knee of the curve. The doubled filter network produces a steep roll-off at high frequencies.

The grey curve is a plot of the voltage across C1, and it can be seen that the strongest feedback occurs around 1 kHz. The response of the filter turns sharply down above 1 kHz. From then on, there is a steady and more rapid fall in response, the rate depending on the number of stages. With two stages in the filter, the roll-off is −12 dB per octave or −40 dB per decade.

This sharpening of the 'knee' is accomplished in a passive filter by using an inductor-capacitor network (p. 57) to produce the resonance. Using an op amp instead provides the same effect but without the disadvantages of inductors.

A highpass version of the second order filter is built by substituting resistors for capacitors and capacitors for resistors. Higher-order filters with even steeper roll-off often consist of two or more second-order filters and possibly a first-order filter cascaded together.

BANDPASS FILTER

A bandpass filter may be constructed by cascading a lowpass filter and a highpass filter (see in Fig 13.8). The cut-off frequency of the lowpass filter is made *higher* than that of the highpass filter so that there is a region of overlap in which both filters pass frequencies at full strength. This area of overlap can be made narrow or broad by choosing suitable cut-off frequencies. Both lower and upper cut-off frequencies are calculated using the equation $f = 1/2\pi RC$.

In the example opposite (Fig 13.7), the cut-off frequency of the lowpass filter is 4.8 kHz, while that of the highpass filter is 1 kHz. The frequency response shows a broad passband from 645 Hz to 7440 Hz, with a central frequency of 2.2 kHz. The bandwidth of the filter is 6795 Hz.

Another approach to bandpass filtering is the **multiple feedback filter** (see Fig 13.11). This has two loops, each providing *negative* feedback.

FIGURE 13.8   A broad-band bandpass filter can be made by cascading a lowpass filter and a highpass filter. This is followed by a non-inverting amplifier to restore the amplitude to a reasonable level.

The result (Fig 13.10) is a passband of narrow bandwidth. The centre frequency is 1.8 kHz. The lower cutoff frequency is 1 kHz and the upper cutoff frequency is 3 kHz. This gives a bandwidth of 2 kHz. The centre frequency (but not the bandwidth or the amplitude) may be tuned by varying the value of R3.

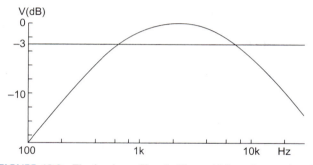

FIGURE 13.9 The bandpass filter in Figure 13.8 produces a broad passband, with response falling off at 6 dB per octave on either side.

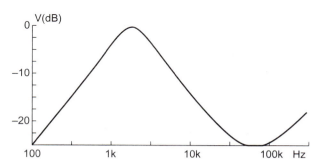

FIGURE 13.10 The bandpass filter in Figure 13.11 produces a narrow passband, with response falling off at 6 dB per octave on either side.

- R1 and C1 form a highpass filter; high frequency signals passing through this are fed back negatively, partly cancelling out the high frequencies in the original signal.
- R2 and C2 form a lowpass filter; low frequency signals passing through this are fed back negatively, partly cancelling out the low frequencies in the original signal.

Only signals in the medium range of frequency are able to pass through the filter at full strength.

FIGURE 13.11 A multiple feedback filter cancels out both high and low frequencies by feeding them back negatively, leaving the middle frequencies unaffected.

The centre frequency of the filter is calculated from the equation:

$$f_0 = \frac{1}{2\pi C} \sqrt{\frac{R_1 + R_3}{R_1 R_2 R_3}}$$

The response curve in Fig 13.10 shows an upturn in the higher frequency range. The is the result of the transition frequency of the op amp. The gain of the op amp is reduced at higher frequencies and therefore the higher frequencies in the signal are not completely cancelled out by the negative feedback.

BANDSTOP FILTERS

Bandstop filters, often known as **notch filters**, are used to remove a limited range of frequencies from a signal. For example, such a filter may be used to remove a 50 Hz mains 'hum' that has been picked up by electromagnetic interference. This may be necessary when measuring small voltage signals from the human body.

There are two approaches to bandstop filtering, both of which show interesting applications of electronic principles. The first circuit (below) depends on one of the basic applications of op amps.

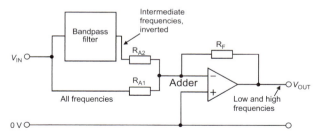

FIGURE 13.12 This bandstop filter uses two op amps, one as part of an active filter, the other as an adder.

The signal is sent to an op amp adder, both directly and indirectly through a bandpass filter. The filter could be a multiple feedback filter (see p. 114). This not only filters the signal, removing low and high frequencies, but also inverts it.

The second op amp mixes the unfiltered signal with the inverted filtered signal. The result is to subtract the medium frequencies from the original signal, creating a notch in the frequency response curve. The depth of the notch depends on the relative values of R_{A1} and R_{A2}.

Another type of bandstop filter depends on an LC resonant network Fig 13.13. This is combined with a resistor and an op amp wired as an inverting amplifier.

Oscillators

An oscillator produces an alternating output voltage of fixed or variable frequency.

The waveform of many types of oscillator is a sinusoid (p. 44) but other waveforms may be generated such as triangular, sawtooth, and square waves. Circuits that produce square waves, often called astable multivibrators, are described on p. 43.

Oscillators may be based on a single transistor but an op amp is preferred for greater reliability. Each of the three oscillators described in this topic operates on a different principle but are all op amp oscillators.

The main requirement is that the op amp should have a full-power bandwidth that includes the intended operating frequency.

PHASE SHIFT OSCILLATOR

All three oscillators rely on **positive feedback** to keep them oscillating. Part of the output of the op amp is fed back to its non-inverting input to maintain the oscillations.

The voltage at the positive terminal of the op amp is held constant at half the supply, and stabilised by the capacitor.

A rising voltage at the output is fed back to the inverting input (−), causing a fall of output. This is negative feedback and the op amp would have a stable output. There is positive feedback too, through the phase shift network.

The network consists of three high-pass filters in series. On p. 48, it is explained that the output from a low-pass filter lags behind the input by up to 90°. Conversely the output of a high-pass filter *leads* input by up to 90°.

Three filters could produce a phase lead of up to 270°. At a particular frequency the lead will be 180°. The signal passing through the phase-shift network is fully *out of phase* with the op amp output. This signal is now amplified and inverted by the op amp, so it is *in phase* with the negative feedback signal. The output from the circuit is unstable and oscillates strongly.

The output waveform is a sinusoid with frequency $f = 1/15.39RC$.

The circuit is a simple one that is best used as a fixed-frequency oscillator. It would be too difficult to have variable resistors or capacitors in the network and to tune them all at the same time.

FIGURE 14.1 The phase shift oscillator consists of two sub-circuits: an op amp adder/inverting amplifier and a phase shift network. The gain of the op amp is set to about 30.

Electronics: Circuits and Systems.

COLPITTS OSCILLATOR

This type of oscillator depends on a resonant network consisting of two capacitors (total series capacitance = C) and an inductor (L) in parallel with them. As shown on p. 79, this L-C network resonates at a frequency, $f = 1/2 \pi \sqrt{(LC)}$.

The op amp is wired as an inverting amplifier with a gain of about 30. Its non-inverting (+) input is held at half the supply voltage (+V/2) by the two 22 kΩ resistors acting as a postential divider.

The LC network is placed in the positive feedback loop of the op amp. At the resonant frequency the output from the op amp makes the network resonate. The tapped point between the capacitors is at +V/2, but the part of the oscillating signal across C2 is fed to the inverting amplifier. It is amplified and keeps the network oscillating strongly.

> **Oscillator in Action**
>
> An animated diagram of a Colpitts oscillator is present on the Companion Site.

When the circuit is first switched on there is no signal, but small random currents in the circuit are amplified. Those at the resonant frequency are amplifed more than others and gradually build up in amplitude until the circuit is oscillating strongly.

If you are trying out this circuit, the capacitors can be 100 nF each. For the inductor, wind two or three turns of ordinary single-stranded connecting wire about 10 mm in diameter. Neither core nor former is needed.

FIGURE 14.2

WIEN BRIDGE OSCILLATOR

This oscillator (diagram below) relies on a pair of LR resonamt networks to shift the output signal by exactly 180° and one particular frequency.

The phase-shifted signal is then fed back to the non-inverting (+) input of the op amp. It is amplified and reinforces the output signal. A strong sinusoidal waveform is generated.

In the Wien network, both capacitors and both resistors are equal in value. The signal frequency is $f = 1/2\pi RC$.

The amplifier has the advantage that it is easily made tunable by using a dual ganged variable resistor for the two resistors in the Wien network.

FIGURE 14.3 The Wien bridge oscillator produces low-distortion sine waves in the audio range.

The gain of the op amp needs to be held very close to 3. Oscillations die away if gain is less than 3, and the op amp becomes saturated if it is more. Gain is set by the variable feedback resistor and the resistance of the filament lamp.

The resistance of the lamp depends on the average current flowing through it, which depends on the amplitude of the signal. With high gain and high current the filament soon heats up and its resistance increases. This increase of resistance decreases the gain. If gain falls, current falls and the filament cools. The resistance of the lamp decreases and gain is reduced. This is a good example of negative feedback leading to stability.

The point at which the gain stabilises is set by the variable resistor. This is adjusted until a strong permanent signal is produced at the output terminal. The amplifier then has the required gain, close to 3.

Things to do

Try breadboarding these oscillators. View their output signals with an oscilloscope, or measure them using a test-meter with a frequency measuring facility.

On the Web

There are thousands of sites describing oscillator circuits. Search with the keyword 'Colpitts oscillator' and visit the sites that describe the many versions of this type of oscillator. You need a design for a practical project; the Web is the place to find it.

Try the same thing with other types of oscillator.

Power Amplifiers

A power amplifier is used to produce a major effect on the surroundings. Examples are power audio amplifiers, producing sound at high volume, or motor control circuits actuating the arm of an industrial robot or aligning the dish of a radio telescope. Other examples include the circuits that produce the dramatic effects of disco lighting. The power of these devices is rated in tens or hundreds of watts, sometimes more. Even a pocket-sized CD player can have a 700 mW output, yet the laser light signal that its head is picking up from the disc is far less powerful than that.

The electrical signals that initiate any of these actions may be of extremely low power. For example, the power output of a microphone or many other types of sensor, or the control outputs from a microcontroller, are usually rated at a few milliwatts. The aim is to amplify the power of the signals from these devices so that they can drive powerful speakers, motors or lamps.

The power at which a device is operating depends on only two quantities, the *voltage* across the device and the *current* flowing through it. The relationship is simple:

$$P = IV$$

The amplifiers that we have described in Topics 8 to 12 have mainly been voltage amplifiers. Some, such as the FET amplifiers, actually produce a current that is proportional to their input voltage but, even then, we usually convert this current to an output voltage by passing it through a resistor. Thus the early stages of amplification are usually voltage amplification. Since P is proportional to V, amplifying the voltage amplitude of a signal amplifies its power by the same amount.

Descriptions of voltage amplifiers often refer to the fact that the currents in amplifiers are small. In many instances the collector or drain current is only 1 mA. There is good reason for this.

As explained in Topic 27, large *currents* through a semiconductor device generate noise. This type of noise is a random signal, which shows up as backgound hissing in an audio circuit. If the audio signal is weak, it may not be possible to pick it out against a noisy background. In other kinds of circuit it may become evident as unpredictable behaviour, making the circuit unreliable.

When a signal has been through several stages of amplification it may have reached an amplitude of a few volts or perhaps a few tens of volts. But, for the reason given above, its current is rated as no more than a few milliamps, perhaps less. Consequently, its power level is low. The next and final stage is to amplify the current.

CURRENT AMPLIFIERS

The two current amplifiers most often used are the common-drain amplifier and the common-collector amplifier. Both of these are voltage follower amplifiers. Their voltage gain is just less than 1 and they both have high current gain. In addition, they have low output resistance, a useful feature when driving high-power devices.

> **Self Test**
>
> What are the alternative names for CD and CC amplifiers ?

The voltage output of these amplifiers is generated by passing a variable current through a resistor, either the drain resistor or the collector resistor. The value of the resistor is chosen so that, when there is no signal, the output voltage sits at half way between the 0 V line and the supply voltage. This gives the output voltage room to swing to maximum extent in either direction without clipping or bottoming.

An amplifier of this type is known as a Class A amplifier. We examine this type in more detail in the next section.

Electronics: Circuits and Systems.

pulses. It then takes longer for the pulse to reach a value that will trigger the thyristor. Firing is delayed until later in the cycle. The plot below shows what happens if R2 is reduced to 560 Ω. The delay at the start of the cycle is now 2.2 ms. This gives a firing angle of $360 \times 2.2/20 = 40°$. The conduction angle is $180 - 40 = 140°$.

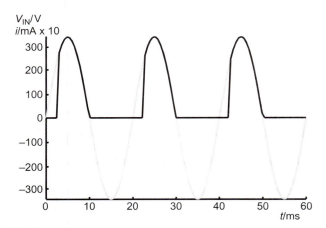

FIGURE 16.6 Reducing the value of R2 increases the delay at the beginning of each cycle, so that the conduction angle is smaller, and the amount of power supplied to the load is reduced.

The value of R2 controls the delay at the beginning of each cycle, which controls the firing angle and conduction angle. The conduction angle determines how much power is delivered to the load during each cycle. The circuit is a **half-wave controlled rectifier** which can be used for controlling the brightness of mains-powered lamps and the speed of mains-powered motors. There is also the fact that the thyristor rectifies the alternating current to produce pulsed direct current, which can be used for driving DC motors. But this is a half-wave circuit so, at the most, the lamp or motor is driven at only half power. This is a basic circuit and several improvements are possible, as will be explained later.

FULL-WAVE CONTROLLED RECTIFIER

There are various ways of using thyristors in full-wave controlled rectifiers. One of these has two thyristors in parallel but with opposite polarity so that one conducts on the positive half cycle and the other on the negative half cycle. They each have a pulse generator synchronised to switch on the thyristor during the appropriate half-cycle. A different circuit, which requires only one thyristor, is illustrated in Figure 16.7.

FIGURE 16.7 A full-wave controlled rectifier has the same circuit as in Figure 16.6, but the power is first rectified by a full-wave diode bridge.

The Companion website has animated diagrams that show how a rectifier operates.

The current is rectified by a conventional diode bridge, and the thyristor acts to turn the current on and off during both positive and negative half-cycles. The thyristor is able to act during both half-cycles because it is located in the rectified part of the circuit so that both half-cycles are positive. This circuit is suitable only for loads that operate on DC.

As indicated in the circuit diagram, the load R_L can be placed in the mains side of the diode bridge instead. The current still has to flow through the thyristor to

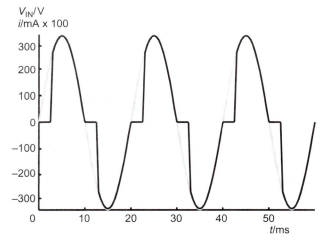

FIGURE 16.8 If the load on a full-wave controlled rectifier is placed on the mains side of the diode bridge, the current through the load (black) is alternating and is controlled during both positive and negative half-cycles.

complete its path through the circuit. The thyristor turns the current on and off during both positive and negative half-cycles, as shown in Figure 16.8.

TRIACS

One of the chief limitations of thyristors is that they pass current in only one direction. At the most, they can operate only at half-power.

If we require power during both half-cycles we need to wire two thyristors in parallel with opposite polarity, or rectify the supply. The triac is a device in which two thyristors are combined in parallel but with opposite polarity in a single block of silicon.

Gating is provided for one of the thyristors, but when the thyristors are combined in this way, the gate switches either one of them, whichever one is capable of conducting at that instant.

Another useful feature is that either positive or negative pulses trigger the device. Like a thyristor, the device switches off whenever the supply voltage falls to zero, and then requires re-triggering. It also switches off if the current through it falls below the minimum holding current.

FIGURE 16.9   Because the triac conducts in both directions, its main terminals are simply named MT1, MT2. The third terminal is the gate.

Although triacs have advantages over thyristors, they have a disadvantage too. Since they conduct on both cycles they have little time to recover as the current through them falls to zero and then increases in the opposite direction.

If the load is inductive, the voltage across the triac may be out of phase with the current through the triac. There may still be a voltage across the triac when the current is zero. This may cause the triac to continue conducting and not switch off.

FIRING PULSES

In the thyristor and triac circuits described so far, we have used a pair of resistors as a potential divider to derive the firing pulse from the mains voltage.

The simple two-resistor system ensures that the pulse occurs at a fixed firing angle. The pulse can be made to occur at any time during the first 90° of the half-cycle (as voltage increases) but the conduction angle must always be between 90° and 180°. If we want a conduction angle between 0° and 90°, we use a capacitor to delay the pulse for more than 90°.

The lamp-dimming circuit below illustrates this technique. The triggering network consists of a resistor and capacitor. If VR1 is increased in value, the charge across the capacitor rises to its maximum later in the cycle.

FIGURE 16.10   A lamp dimming circuit using a triac, with a resistor-capacitor network to trigger the triac at any stage during the cycle.

The plot below shows firing at 126° and 306°. The triac is triggered by a positive pulse during the positive half-cycle and by a negative pulse during the negative half-cycle.

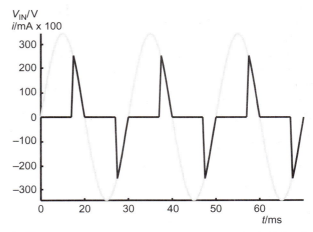

FIGURE 16.11   The triac circuit dims the lamp completely if the triac is triggered late in each half-cycle.

Although the resistors and capacitors in triggering networks pass very small currents, and so are not wasteful of power, there is always the risk of their eventually breaking down. They are directly

As an example, we could supply power to a motor for five half-cycles, then cut off the power for the same number of half-cycles. This supplies power for half the time. Power is interrupted for five half-cycles (for 0.1 s on a 50 Hz mains supply) but the inertia of the motor and the mechanism it is driving keeps it rotating during the off periods. An electric heater may also be controlled by this technique because interruptions of a fraction of a second or even several seconds makes little difference to the temperature of the heating element.

On the other hand, this form of control is not suitable for controlling the brightness of lamps because it makes them flicker noticeably.

Integral cycle control relies on being able to detect the instant at which the alternating supply has zero value and is changing direction.

This is the zero-crossing point at the beginning of every half-cycle. A zero-crossing detecting circuit can be built from discrete components but special integrated circuits are more often used. Some triac opto-couplers have a built-in zero-crossing detector that allows the triac to be switched on only at the beginning of the half-cycle.

A typical example of a zero-crossing switch IC is the CA3059, which can be used to control either thyristors or triacs. The diagram below shows it being used to provide on/off control of a triac. The integrated circuit contains four main sub-circuits:

- A regulated 6.5 V DC power supply for the internal circuits. This has an outlet at pin 2 for supplying external control circuits.
- Zero-crossing detector, taking its input from the mains supply at pin 5.
- High-gain differential amplifier, with its inputs at pins 9 and 13. Depending on the application, this can accept inputs from:
  - a switch (as in the diagram), for simple on/off control.
  - a sensing circuit based on a device such as a thermistor or light-dependent resistor, for automatic control of a heater or lamp.
  - a transistor or opto-isolator, for control by automatic circuits, including computers and microcontrollers.
  - an astable for burst-fire control.
- Triac gate drive with output at pin 4. It produces a pulse when the mains voltage crosses zero volts. Pulses are inhibited if the pin 9 input of the differential amplifier is positive of the other input at pin 13. Pulses may also be inhibited by a signal applied to the 'inhibit' input at pin 1 and the 'fail-safe' input at pin 14. The drive can be triggered by an external input at pin 6.

As can be seen from the diagram, the CA3059 needs very few external components. These comprise:

- The triac, with its load R_L.
- A snubber network connected across the terminals of the triac (not shown in the figure).
- A dropper resistor R_3, which should be a high-wattage type (about 5 W).
- Smoothing capacitor C_1 for the 6.5 V supply.
- A control circuit.

In the figure, the control circuit consists of S1 and a pair of resistors. Pin 9 is connected to pins 10 and 11 and this puts it at mid-rail voltage (approx 3 V). When S1 is open, R2 pulls pin 13 down toward 0 V. The gate drive circuit is inhibited because pin 13 is negative of pin 9 and the triac does not fire.

When S1 is closed, pin 13 is raised to just over 4 V.

This makes it more positive than pin 9 and enabling the gate drive to produce pulses. The triac is

FIGURE 16.17 The integrated circuit is a zero-crossing switch, used to minimise RFI and to interface the triac to the control circuitry, represented here by a simple on/off switch.

fired at the beginning of every half-cycle and full power is applied to the load.

ACTIVITY — THYRISTOR ACTION

Assemble the circuit shown below. Apply the 12 V DC supply from a bench PSU.

FIGURE 16.18 This circuit is used for investigating the action of a thyristor. Note that the supply is 12 V DC.

Record and explain what happens when:

- power is first applied.
- S1 is pressed and held.
- S1 is released.
- S2 is pressed and held.
- S2 is released.

Measure the current flowing to the gate when S1 is pressed and held. Measure the current flowing through the lamp when it is lit. Add a 22 kΩ variable resistor VR1 to the circuit. Begin with VR1 set to its minimum resistance so that maximum current flows. The lamp lights when S1 is pressed. Measure the current through it.

Repeat with VR1 set to various higher resistances and find the minimum holding current of the thyristor. The lamp may not light when current is small but the meter shows if a current is flowing.

ACTIVITY — CONTROLLED RECTIFIER

Set up the half-wave and full-wave circuits illustrated in Figures 16.4 and 16.7. Note the remarks in the box headed 'Operating voltage' (p. 133) and use a low-voltage AC supply, NOT the mains. Use an oscilloscope to monitor the input and output waveforms.

EXTENSION BOX 18 Phototransistors

When light falls on the base of a transistor, some of the electrons in the lattice gain energy and escape. This provides electron-hole pairs, and is equivalent to supplying a base current to the transistor.

If the collector is positive of the emitter, the usual transistor action then occurs and a relatively large collector current flows through the transistor.

Transistors are usually enclosed in light-proof packages so that this effect can not occur, but a phototransistor is enclosed in a transparent plastic package or in a metal package with a transparent window.

Many types of phototransistor have no base terminal, as the device will work without it.

EXTENSION BOX 19 Opto Devices

Opto devices have an LED and an amplifying semiconductor device enclosed in a light-proof package. Their function is to pass a signal from one circuit to another. Opto **couplers** are used when, for *electronics* reasons, it is not suitable to connect the circuits directly (perhaps voltage levels are not compatible, or we want the two 'ground' rails to be isolated from each other to prevent interference being carried across). Opto **isolators** are specially designed to withstand a large voltage difference (usually 1000 V or more) between the two sides and are used for *safety* reasons, when a high-power circuit is to be controlled by a low-power circuit.

PROBLEMS ON THYRISTORS AND TRIACS

1 A thyristor has its anode several volts positive of its cathode. A positive pulse is applied to its gate. Describe what happens next.

2 Once a thyristor is conducting, what can we do to stop it?

3 Describe how a thermistor and a thyristor may be used to control an electric heating element powered by the mains.

4 Explain why it is preferable to use a thyristor rather than a transistor for controlling large currents.

5 Explain what is meant by *firing angle* and *conduction angle*.

6 Draw a circuit diagram of a thyristor half-wave controlled rectifier and explain its action.

7 Draw a diagram of a simple phase-control triac circuit that could be used in a mains-powered lamp-dimming circuit.

8 List the methods used to isolate a triac circuit from its control circuit.

9 Explain the difference between an opto isolator and an opto coupler.

10 Explain the possible causes of false triggering in a triac power circuit and what can be done to prevent it.

11 Explain the differences between phase control and integral cycle control, and their advantages and disadvantages.

12 Describe the action of the CA3059 zero crossing switch IC and some of its applications.

MULTIPLE CHOICE QUESTIONS

1 A thyristor is triggered by:
   A a negative pulse applied to the gate.
   B a negative pulse applied to the anode.
   C a positive pulse applied to the cathode.
   D a positive pulse applied to the gate.

2 Starting from the anode, the semiconductor layers in a thyristor are:
   A pnpn.
   B npn.
   C pnp.
   D npnp.

3 Triacs are triggered by:
   A positive or negative pulses.
   B positive pulses only.
   C negative pulses only.
   D pulses produced as the mains voltage crosses zero.

4 A snubber network is used to reduce:
   A overheating of the triac.
   B RFI.
   C the length of the trigger pulse.
   D false triggering.

5 Once a thyristor has been turned on, it conducts until:
   A the triggering pulse ends.
   B the current through it falls below the holding current.
   C the anode is made positive of the cathode.
   D the gate is made negative of the cathode.

Power Supplies

Electronic circuits normally operate at low voltages and on direct current (DC). A typical circuit operates on 6 V or 9 V DC, but some can work on as little as 1 V. Often this low voltage supply is provided by a battery. A battery is ideal for portable equipment such as a mobile phone or a digital camera. But batteries are fairly expensive to replace, especially if the equipment is to be operated for long periods or requires a large current. In such cases, we prefer to use a power supply that takes power from the mains and converts it to a low voltage DC supply suitable for electronic circuits. DC power supplies are found in many different kinds of electronic equipment from TV receivers and computers to washing machines and cash registers.

This topic illustrates three ways of building a power supply unit (PSU) suitable for running electronic circuits from the mains. This topic describes the simplest of the circuits, which has three stages:

FIGURE 17.1

The functions of these stages are:

- **Transformer:** transforms mains power (230 V AC) into a lower AC voltage.
- **Rectifier:** converts low AC voltage into low DC voltage.
- **Smoothing capacitor:** smooths the DC output of the rectifier.

The **transformer** (T1) is of the step-down type wound on a soft-iron laminated core. The primary winding is rated to be connected to the live and neutral lines of the 230 V AC mains. The secondary winding has fewer turns than the primary winding, giving a turns ratio of less than 1. The amplitude of the transformed AC is typically as low as 12 V AC. The wire of the secondary winding should be rated to supply the current that it is intended to draw from the PSU.

There should be a **switch** (S1) between the mains plug and the transformer. This should be the double-pole single-throw type so that, when it is off, the power supply circuit is completely disconnected from the mains. The case should be made of metal or stout plastic able to withstand rough treatment. The Earth line of the mains is connected to a point on the chassis of the PSU, and to the case, if the case is made of metal.

The **diode** is a rectifier diode, designed to pass a relatively large current without damage. The photo shows types of rectifier diode suitable for use in a PSU.

FIGURE 17.3

FIGURE 17.2 The simplest PSU has a half-wave rectifier, based on a single diode.

Electronics: Circuits and Systems.

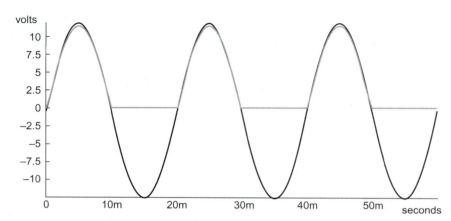

FIGURE 17.4 The input to the rectifier circuit is low-voltage AC at 50 Hz, plotted here with a black line. The output (grey line) is pulsed DC because the diode conducts only on the positive half-cycles.

RECTIFICATION

The output from the transformer is a sinusoid. The diode is able to conduct only during the positive half-cycle of the sinusoid.

The waveform consists of a series of positive pulses occuring 50 times per second. This waveform is known as **pulsed DC**. Because of the forward voltage drop across the diode, the amplitude of the pulses is 0.7 V less than that of the transformed AC. The circuit produces current only during the positive half-cycles, so it is called a **half-wave rectifier**. It is inefficient because the diode is conducting for only half of the time.

SMOOTHING

For applications such as driving a small DC motor, the pulsed DC can be used directly from the rectifier.

If only a small current is required, the waveform can be smoothed by connecting a capacitor (C1) across the output terminals. An electrolytic capacitor of large value (for example, 1000 μF) is needed.

OUTPUT

To protect the transformer and diode against over-loading or short-circuiting, include a fuse on the output side. This is not shown in the diagram.

With no load, the output is several volts higher than expected from the turns ratio of the transformer.

Voltage drops to the expected value when a small load is connected.

This PSU is **unregulated**, so the output voltage decreases if the load draws increasing amounts of current.

> **Rectifiers**
>
> The Companion website has animated diagrams that show how rectifiers operate.

FULL-WAVE RECTIFIED AND ZENER REGULATED SUPPLY

This PSU is slightly more complex and expensive than the one just described, but is more efficient and produces a regulated output voltage. It comprises four stages:

- **Transformer:** As in the unregulated supply, this transforms mains power into a lower AC voltage.
- **Rectifier:** converts low AC voltage into low DC voltage.
- **Smoothing capacitor:** smooths the DC output of the rectifier.
- **Zener regulator:** holds the output voltage steady in spite of variations in the amount of current drawn by the load.

The two new improvements in this PSU are the rectifier bridge of four diodes and the Zener regulator circuit. The system diagram now has four stages:

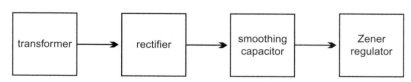

FIGURE 17.5

FULL-WAVE RECTIFIER

The four diodes D1 − D4 are connected to form a rectifying **bridge**. This receives AC from the transformer.

During the positive half-cycle, terminal A of T1 is positive of terminal B. Current flows from A, through D1, and through R1 to the load. It passes through the load and returns through D2 to B.

During the negative half-cycle, terminal B is positive of terminal A. Current flows from B, through D3, and through R1 to the load. It passes through the load and returns through D4 to A.

Summarising, D1 and D2 conduct on the positive half-cycle, while D3 and D4 conduct on the negative half cycle.

During both half-cycles the current passes through two diodes, so there are two diode drops, totalling 1.4 V

There is conduction during *both* half-cycles, which is why this is called a full-wave rectifier. Given a 50 Hz supply, the output from the diode bridge is pulsed DC, frequency 100 Hz.

Because of forward voltage drop, the amplitude of the pulsed DC is 1.4 V less than that of the AC produced by the transformer.

SMOOTHING

The diagram above (Figure 17.4) shows the pulsed DC output from the rectifier stage. The diagram below shows the effect of the smoothing capacitor C1. The capacitor charges at each pulse. The voltage across the capacitor reaches its maximum level, which equals the peak voltage from the rectifier. Then, as the output voltage from the rectifier falls, the capacitor retains all or most of its charge.

If a load is connected, some of the charge is conducted away from the capacitor. The voltage across the capacitor falls.

In a properly designed PSU, the fall in voltage is small. Before the voltage has had time to fall far, the next pulse arrives and the capacitor is recharged to the peak level. The effect of charging and partial discharging is a slight **ripple** on the DC level. The graph in Figure 17.8 shows this ripple.

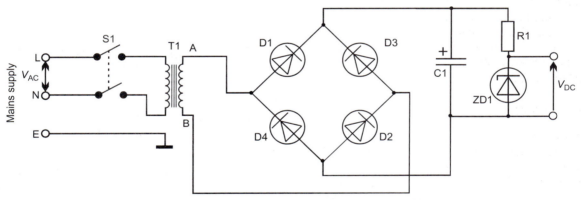

FIGURE 17.6 The diode rectifier bridge and the Zener diode give improved performance.

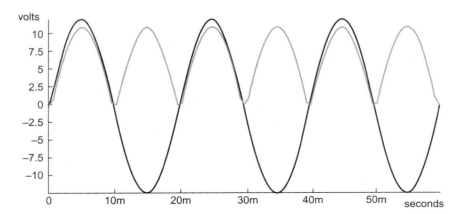

FIGURE 17.7 Without the electrolytic capacitor shown in the diagram, the output (grey) waveforms of a full-wave rectifier is pulsed DC with a frequency double that of the input (black). The amplitude is 1.4 V less than the input because of the voltage drop across two diodes.

The depth of the ripple depends on the capacity of C1 and the amount of current drawn by the load. The greater the current, the deeper the ripple.

Ripple may be minimised by using a large-value capacitor (1000 μF or more) for C1, so that the pd across it falls more slowly.

CALCULATING RIPPLE

It can be seen from the graph below that ripple is a more-or-less triangular waveform superimposed on the steady DC output.

There is a rapid rise as the charge on the capacitor is boosted to the peak voltage, followed by a slow fall as charge is conducted away through the load. To simplify the calculations, we will assume that the rise is almost instantaneous and that the fall lasts for one complete cycle of the waveform. To begin with, we will consider what happens with a half-wave rectifier and no subsequent voltage regulation, as in the circuit in Fig. 17.2.

If I is the mean load current and T is the period of the waveform, then the change in charge on the capacitor is:

$$q = IT$$

The *fall* in the voltage across the capacitor is:

$$v = IT/C$$

Given that $T = 1/f$, where f is the frequency, this equation can be rewritten:

$$v = I/Cf$$

We now consider what happens in different types of power circuit. In the half-wave rectifier power supply, the output voltage falls from V_p (peak voltage) to

$V_p - v$ at an approximately constant rate. This being so, the mean output voltage is:

$$V_{mean} = V_p - \tfrac{1}{2} \times \frac{I}{Cf} = V_p - \frac{I}{2Cf}$$

The peak voltage is reduced by an amount that is proportional to the current being drawn and is inversely proportional to the capacitance and frequency. This explains why the smoothing capacitor should have as large a capacitance as practicable.

Self Test

The output of a half-wave rectifier has peak voltage 6.5 V. The smoothing capacitor is 2200 μF, and the frequency is 50 Hz. What is the mean output voltage when the current drawn is (a) 25 mA, and (b) 120 mA?

With a full-wave rectifier, the ripple frequency is double that of the AC frequency. If f is the AC frequency, the fall in output voltage is:

$$v = \frac{I}{2Cf}$$

This results in a mean output voltage of:

$$V_{mean} = V_p - \frac{1}{4Cf}$$

This shows that the fall in voltage for a given load is halved when a full-wave rectifier is used.

The descriptions above refer to unregulated power supplies. The effect of ripple is much reduced in stabilised supplies, such as the Zener stabilised supply of Fig. 17.9 and the regulator stabilised supply of Fig. 17.10. The variations in voltage due to ripple are largely eliminated by the stabilising action of the Zener diode or the regulator.

FIGURE 17.8 Adding a capacitor to the output side of the rectifier produces DC with a small 100 Hz ripple.

REGULATION

The output voltage of this PSU is held steady under varying load by using a regulator circuit. This is based on a **Zener diode**. The Zener diode needs about 5 mA or more passing through it to maintain its Zener action. This is the reason for the resistor R1. The value of this is chosen to allow at least 5 mA to flow through the diode, even when the load is drawing its maximum current.

FIGURE 17.9 When the load is drawing the maximum current, there must be at least 5 mA passing through the Zener diode.

In the diagram above, the load is drawing its maximum current, I_{MAX}. The current through the Zener diode is I_Z. The current through R1 is $I_{MAX} + I_Z$.

The pd across the Zener diode is its Zener voltage, V_Z. If V_{IN} is the smoothed voltage across the capacitor of the PSU, the pd across R1 is $V_{IN} - V_Z$. By Ohm's Law, the minimum value of R1 is:

$$R_1 = (V_{IN} - V_Z)/(I_{MAX} + I_Z)$$

Example

A PSU has a smoothed voltage of 10.8 V. The Zener voltage is 5.6 V. The maximum current that the load may take is 250 mA. Calculate the value of R1.

$$R1 = (10.8 - 5.6)/(0.25 + 0.005) = 5.2/0.255$$

$$= 20.4 \ \Omega$$

Use the next larger E24 value, 22 Ω.

ZENER POWER RATING

It is essential for the Zener diode to be able to carry the full current $I_{MAX} + I_Z$ when there is no load connected to the PSU.

Under this condition, the pd across the Zener diode is V_Z and the current through it is $I_{MAX} + I_Z$. The power dissipated in the diode is then:

$$P = (I_{MAX} + I_Z) \times V_Z$$

Example

Find the required power rating of the Zener diode used in the previous example.

$$P = (0.25 + 0.005) \times 5.6 = 1.428 \ W$$

Zener diodes are made in standard power ratings of 0.4 W, 1.3 W and 5 W. Select a 5 W diode.

RESISTOR POWER RATING

The current through the resistor is $(I_{MAX} + I_Z)$ and the pd across it is $(V_{IN} - V_Z)$. Therefore, the power dissipated in the resistor is:

$$P = (I_{MAX} + I_Z) \times (V_{IN} - V_Z)$$

Example

Calculate the required rating for the resistor in the previous example.

$$P = (0.25 + 0.005) \times (10.8 - 5.6) = 1.326 \ W$$

A 2 W resistor should be used.

LOAD REGULATION

Load regulation is expressed as the percentage change in output voltage as the current drawn is varied from zero up to the maximum for which the supply is designed.

Example

Under no load, the mean output voltage of a supply is 12 V. When the load is drawing the maximum current of 2.5 A, the output voltage falls to 11.1 V. What is the load regulation?

Change in output = 12 − 11.1 = 0.9 V.
Load regulation is 0.9/12 × 100 = 7.5%.

HIGH CURRENT

In the basic Zener circuit, the diode must be rated to carry the maximum output current, plus 5 mA. This means that an expensive power Zener may be required. Further, when the maximum current is not being drawn by the load, a large current is passed through the Zener, which is wasteful.

Self Test

A Zener-regulated mains PSU using a full-wave rectifier is to provide a current of up to 400 mA at 7.5 V. The transfomer is rated to produce an output of 12.5 V AC at 1 A.
a   What is the current rating of the rectifier diodes?
b   What is the voltage of the output from the rectifier?
c   Specify what type of Zener diode you would select for this PSU.
d   What is a suitable resistance and power rating for the dropper resistor?

The diagram below shows an alternative Zener regulator circuit, in which Q1 is connected as an emitter follower (pp. 75–76).

FIGURE 17.10   With the MJE3055, this circuit can deliver up to 10 A, regulated.

Input v_{IN} comes from a rectifier or a battery. The Zener voltage V_Z is 0.7 V more than the required output voltage. A low-power (400 mA) Zener is suitable. R1 is chosen so that the current through D1 is more than 5 mA, say 10 mA. The Zener holds the base of Q1 steady at V_Z, and the base-emitter voltage drop produces $(V_Z - 0.7)$ V at the output.

FULL-WAVE RECTIFIED AND IC REGULATED SUPPLY

This circuit improves on the previous two power supply circuits by using:

- an encapsulated diode bridge rectifier, and
- an integrated circuit voltage regulator.

FIGURE 17.11   In this circuit the Zener diode regulator is replaced by an IC voltage regulator.

DIODE BRIDGE

In an encapsulated diode bridge the four power diodes are connected to form a rectifier bridge, as on Fig. 17.6.

FIGURE 17.12   A bridge rectifier makes circuit construction simpler.

They are sealed in a capsule with four terminal wires. Two of the wires ('AC' symbols) receive the transformed AC direct from the transformer and the rectified DC output is obtained from the other two wires (+ and −). The bridge in the photo is rated at 6 A.

The diode bridge works the same way as a bridge built from four individual diodes but an encapsulated bridge is easier to mount and simpler to connect.

REGULATOR IC

Many types of regulator IC are available, but one of the most popular is the three-terminal regulator. The 7805 IC is an example of this type.

The 7805 is a member of the '78XX' family, which includes regulators of different power ratings, producing various fixed output voltages. The output is indicated by the final two digits of the type number. The 7805, for example, produces a fixed output voltage of 5 V. Other ICs are available for variable voltage regulation (*see* overleaf).

The precision of a typical voltage regulator is ±4%, which is held in spite of variations in the output current and the supply voltage (*see* next section). This is better performance than can be obtained from a Zener regulator under similarly wide-ranging conditions. The supply voltage should be about 2.5 V higher than the output voltage. This allows good regulation

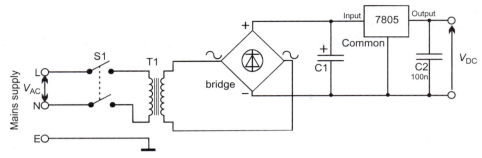

FIGURE 17.13   The classic voltage regulator IC is a three-terminal device.

without undue dissipation of power within the regulator IC. The maximum allowable supply voltage is 30 to 40 V.

The 7805 is rated to produce a current of 1 A. Similar regulators are available for currents of 100 mA and 500 mA. A heat sink may be needed with the higher current ratings. The IC has a metal tag with mounting hole for attaching this.

LOAD AND LINE REGULATION

Load regulation was defined earlier. For a typical IC regulator, the load regulation is about 0.1%. This compares with a load regulation of several percent for a basic Zener diode regulator.

Line regulation is a measure of the change in output voltage resulting from a change in input voltage. It is defined as:

$$\frac{\text{change in output voltage}}{\text{change in input voltage}} \times 100\%$$

Again, IC regulators have superior performance. A typical IC regulator has a line regulation of 0.01%, compared with several percent for a basic Zener regulator.

SAFETY FEATURES

All IC regulators have a **current limiting** feature. If excess current is drawn from the regulator, perhaps because its output is short-circuited, the output voltage is automatically limited. This limits the size of the current that can be drawn from the device and so protects it and the load from damage.

Another safety feature is **thermal shut-down**. This comes into effect if the device becomes overheated. This too can be the result of short-circuiting the output, or perhaps connecting a unsuitably low resistance across the output. When the IC becomes too hot, its output is cut to zero.

ADJUSTABLE REGULATOR

The LM317 is typical of the 3-terminal adjustable regulators.

The regulator can accept a maximum voltage input of 40 V. The output voltage is adjustable from 1.2 V to 32 V. The input voltage must be at least 3 V greater than the intended output voltage. The output voltage is set by the value of the resistor R in the diagram below.

FIGURE 17.14   The diodes protect the regulator against short-circuiting either the input or output terminal to the 0 V rail.

The value of R is calculated from:

$$R = 96 \times v_{\text{OUT}} - 120$$

Example

In an LM317 regulator circuit, what value of R is required to produce an output voltage of 10.5 V?

$$R = 96 \times 10.5 - 120 = 888 \; \Omega$$

The nearest E24 value is 910 Ω. If this is not close enough, use 820 Ω and 68 Ω in series, preferably with 1% tolerance.

ACTIVITIES — POWER SUPPLIES

Investigate the action of two of the power supply circuits and write a report including diagrams of the two circuits and comparing their advantages and disadvantages. Use a signal generator **(NOT THE MAINS)** to provide a low-voltage AC input. A signal of 10 V AC, running at 50 Hz, is suitable. Design the circuit to produce an output of 5 V DC.

Monitor the waveforms present at various stages, including amplitude and shape. Try the effect of loads of various sizes and their effect on smoothing and ripple. Estimate the load and line regulation of the circuits.

EXTENSION BOX 20   Zener Diodes

A typical diode, such as a rectifier diode, can withstand a reverse voltage of several hundred volts before breaking down. Breakdown results in irreversible damage to the diode.

A Zener diode breaks down at a precisely known reverse voltage that is often only a few volts. At this **Zener voltage**, V_Z, current flows freely through the diode without damaging it. The Zener voltage is determined during manufacture. Diodes are available with specfied Zener voltages ranging from 2.0 V up to 100 V or more.

As the power supply circuit in Figure 17.6 illustrates, a Zener diode makes a simple voltage regulator. However, the Zener voltage is not perfectly constant. It rises slightly as the current increases. It also depends on temperature. For these reasons, a Zener regulator has relatively low precision.

QUESTIONS ON POWER SUPPLIES

1 Describe how a half-wave rectifier works and sketch the waveform that it produces. Compare its voltage amplitude with that of the AC it rectifies.

2 Why is a full-wave rectifier more efficient that a half-wave rectifier? Sketch the waveform of the rectified current. Compare its voltage amplitude with that of the AC it rectifies.

3 Describe the action of the smoothing capacitor in a full-wave power supply.

4 A power supply unit has a smoothed voltage of 8 V. The Zener voltage is 3.9 V. The PSU is rated to provide up to 400 mA. Calculate the required resistance and power rating of the dropping resistor.

5 In the previous question, what should be the power rating of the Zener diode?

6 What are the advantages of an IC voltage regulator when compared with regulation by a Zener diode?

7 List and discuss the safety features and precautions that should be considered when building a mains-powered PSU.

8 Design a regulated power supply to provide 1 A at 12 V DC with 4% precision.

9 Design a mains PSU for powering a 6 V DC electric motor, at a maximum of 4 A.

10 Design a power supply for powering a logic circuit that requires 5 V $\pm$ 0.5 V, at 3 A. (This is typical of the requirements for a TTL circuit).

11 Explain the cause of ripple on the output of a full-wave rectified and smoothed power supply circuit. Given: peak voltage = 15 V, the current = 2 A, smoothing capacitor = 4700 μF, and frequency of the rectified AC = 100 Hz, what is the mean output voltage? What can be done to increase this voltage without altering the load current?

12 A simple Zener regulator is loaded with different resistors. The table shows the corresponding Zener voltage. If the circuit is rated to supply up to 240 mA, what is its load regulation?

Load (Ω)	V_Z
15	4.05
20	4.76
25	5.30
30	5.70
35	5.70
No load	5.70

13 The table shows the output voltage of a simple Zener regulator as the supply voltage in increased from 8 V to 18 V. What is the line regulation of this circuit?

V_S	V_{OUT}	V_S	V_{OUT}
8	4.62	14	5.79
10	5.66	16	5.84
12	5.73	18	5.90

Logical Circuits

Logical circuits perform **logical operations**. There are three basic logical operations:

- NOT, sometimes called INVERT.
- AND
- OR

Logical operations are performed by circuit units called **gates**. Because a typical logical circuit needs several, or maybe many, gates they are made as integrated circuits. The drawing below shows the symbol for an AND gate and a typical logic IC containing four AND gates. The gates are separate but they share the power supply pins (0 V and +5 V).

2-input AND gate

7408   74LS08

FIGURE 18.1 The 7408 IC contains four 2-input AND gates in a 14-pin package. A circular dimple in the top surface indicates which is pin 1. Or there may be a notch in the end that is nearer to pins 1 and 14.

AND GATE OPERATION

The drawing at top right shows a circuit for demonstrating the action of a 2-input AND gate. The IC has three other AND gates, but these are not being used, so their inputs are connected to the 0 V rail or the positive rail.

In the drawing we see a commonly used way of providing input to a logic circuit. The inputs of the gate are held at 0 V by the pull-down resistors R1 and R2. We say that the inputs are 'low'. When a button is pressed, the corresponding input goes to +6 V. We say that the input is 'high'. These buttons are **interfaces** between the gate and the human user.

Unused pins 4, 5, 9, 10, 12, 13 to 0 V or +6 V.

FIGURE 18.2 A circuit for demonstrating the action of a 2-input AND gate. Manual input is by the two push-buttons with pull-down resistors. The state of the output is indicated by an LED.

The other interface between gate and the user is an LED. This has a series resistor to limit the current and protect the gate. The LED is not lit when the output of the gate is at 0 V (= 'low'). The LED comes on when the output of the gate goes to +6 V (= 'high').

> **Gate**
>
> A logic gate has nothing to do with the gate of an FET or a thyristor.

When connected as above, the gate acts as a **source** of current that turns the LED on when the output is high. A gate can also act as a **sink**, as shown in Figure 18.3.

The behaviour of the gate is investigated by pressing various combinations of the two buttons S1 and S2. Because there are two inputs to this gate, there are four possible combinations. We record the output of the gate for each combination. This table summarises the results:

Inputs		Output
B	A	Z
0	0	0
0	1	0
1	0	0
1	1	1

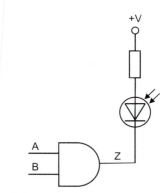

FIGURE 18.3 If the gate is wired as a current sink, the LED comes on whenever the output of the gate is low. At other times the LED is off.

A '0' in the input columns indicates that the button is not pressed, so the input is low. A '1' indicates that the button is pressed, so the input is high.

A '0' in the output column indicates that the gate output is low, and the LED is off (the gate is wired as a source). A '1' indicates that the output is high and the LED is on.

This table of results is identical with the truth table for the AND operator. In words, the output is high only if inputs A AND B are high. If an AND gate has three or more inputs, the output is high only if *all* inputs are high.

Another way of stating this result uses **Boolean algebra**:

$$Z = A \bullet B$$

This statement uses the AND operator, '•'. It tells us that output Z is the result of ANDing input A with B.

A similar statement covers the result from a four-input AND gate:

$$Z = A \bullet B \bullet C \bullet D$$

The output is high only if all four inputs are high.

In Boolean algebra, we may omit the '•' operator and write instead: $Z = AB$, or $Z = ABCD$.

OR GATES

To investigate an OR gate, we may use the same circuit as in Figure 18.2, but with a 74HC32 IC instead.

This contains four 2-input OR gates. The symbol for a 2-input OR gate is drawn below:

FIGURE 18.4

The behaviour of the OR gate is summarised in this table:

Inputs		Output
B	A	Z
0	0	0
0	1	1
1	0	1
1	1	1

This is the truth table for the OR operation. Compare this with the AND truth table in the left column. The inputs are the same as before, but the outputs are different. In words, the output is high when either A OR B OR both are high.

The Boolean expression for this result is:

$$Z = A + B$$

This statement uses the OR operator, '+', which is the same as the 'plus' operator used in arithmetic. This operator must not be omitted from Boolean equations.

OR gates are available with three or more inputs. The result from a four-input OR gate is:

$$Z = A + B + C + D$$

NOT GATE

The NOT gate can have only one input terminal and its output is always the inverse (or opposite) of its input. The 74HC04 IC contains six NOT gates. These can be investigated as described above.

This is the symbol for a NOT gate:

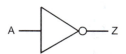

FIGURE 18.5

The small circle at the output indicates that the output is inverted. The truth table of the gate is simple:

Input	Output
A	Z
0	1
1	0

In symbols, the operation of the gate is:

$$Z = \bar{A}$$

In Boolean algebra, a bar over a variable is used to show that it is inverted.

NAND AND NOR GATES

The three basic gates that we have already described can be connected together in a wide variety of ways to perform more complex logical operations. We shall look at some of these in following topics. For the moment, we will describe one of the most useful combinations of gates, an AND gate that feeds its output to a NOT gate. This NOT-AND, or NAND gate is illustrated below:

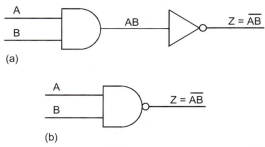

(a)

(b)

FIGURE 18.6   A NAND gate is the equivalent of an AND gate followed by a NOT gate. NAND gates are also made with three or more inputs.

The truth table for the two-input NAND gate may be obtained by writing the table for AND and then inverting the output:

Inputs		A • B	$Z = \overline{A \cdot B}$
B	A		
0	0	0	1
0	1	0	1
1	0	0	1
1	1	1	0

The third column of the table holds A AND B. This is inverted in the fourth column, as indicated by the line drawn over 'A • B'.

The NOR gate is equivalent to an OR gate followed by a NOT gate. In this table we look at a 3-input NOR gate:

Inputs			A+B+C	$Z = \overline{A+B+C}$
C	B	A		
0	0	0	0	1
0	0	1	1	0
0	1	0	1	0
0	1	1	1	0
1	0	0	1	0
1	0	1	1	0
1	1	0	1	0
1	1	1	1	0

It is essential to include all possible combination of inputs in the first three columns. Enter the inputs as the digits of the eight binary 3-digit numbers, 000 to 111. This rule applies to 2-input gates, as can be seen opposite. It also applies to larger numbers of inputs. For instance, with a 4-input gate the truth table has 16 rows, running from 0000 to 1111.

The drawing below illustrates the 3-input NOR gate.

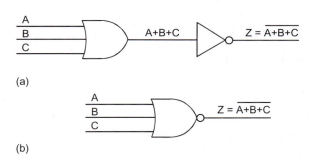

(a)

(b)

FIGURE 18.7   A 3-input NOR gate (b) is equivalent to a 3 -input OR gate followed by a NOT gate (a).

EXCLUSIVE-OR GATE

An exclusive-OR, or EX-OR gate has two inputs A and B. Its output is '1' when A or B *but not both*, are '1'. This is the truth table:

Inputs		$Z = A \oplus B$
B	A	
0	0	0
0	1	1
1	0	1
1	1	0

Note the symbol $\oplus$ for the exclusive-OR operation. Except for the last line, the table is identical to the table for OR. The symbol for an EX-OR gate is:

FIGURE 18.8

The EX-OR gate is sometimes known as the 'same or different' gate. Its output is '0' when its inputs are the same and '1' when they are different. This makes the gate useful for comparing two logical quantities.

The exclusive-NOR gate is equivalent to an EX-OR gate followed by a NOT gate. It has only two inputs. Its output is the inverse of EX-OR.

LOGIC FAMILIES

There are three main logic families:

- Transistor-transistor logic (TTL).
- Complementary MOSFET logic (CMOS).
- Emitter coupled logic (ECL)

Each of these families has its own versions of the standard logic gates such as NAND and NOR. Each family includes a range of ICs with more complex functions such as flip-flops, adders, counters, and display drivers. The families differ in the ways the basic gates are built, and this gives rise to family differences in operating conditions and performance. We will look at each family in turn.

TRANSISTOR-TRANSISTOR LOGIC

This was the first family to become widely used and set the standards for logic circuits for many years.

The devices in the original TTL family were all given type numbers beginning with '74'. This was followed by two or three digits to distinguish the types. For example 7400 is a quadruple 2-input NAND gate, and 7493 is a 4-bit counter/divider. To refer to the family itself we substitute 'XX' for the last two or three digits.

The original 74XX series is now virtually obsolete, being used mostly for repairing old equipment, and is becoming much more expensive than it used to be.

There are over 20 more recent versions of the series differing in various features, such as power, speed, immunity from noise, EMI reduction, and supply voltage. Some of the newer versions operate with low power on a 3.3 V supply, which makes them suitable for battery-powered portable equipment, as well as being more economical of power generally.

One of the most popular series is the low power Schottky series, with device numbers 74LSXX. This has a faster operating time than standard TTL. Both standard and 74LSXX ICs require a regulated power supply of 5 V $\pm$ 0.25 V.

The output stage of TTL gates has two transistors operating in opposition. One switches the output to high, the other switches it to low. In operation, there is a changeover stage during which *both* of the transistors are switched on. This causes a large current to flow as the gate changes state. This may overload the supply line, causing a fall in voltage which may interfere with the action of nearby ICs. For this reason the supply to TTL ICs must be decoupled by wiring capacitors between the supply lines. Decoupling usually requires a 100 nF disc ceramic capacitor for every 5 ICs.

Certain TTL logic devices are made with an **open collector output**. This has a transistor with its base and emitter connected to the gate circuit but its collector left unconnected. The output must have a pull-up resistor connected to it (1 kΩ is often suitable).

FIGURE 18.9   The output of TTL gates that have an open collector output can be pulled up to voltages higher than the standard 5 V operating voltage.

The resistor is connected to a voltage higher than the standard 5 V operating voltage. Depending on the type of gate, the higher voltage can be as much as 15 V or 30 V. Provided that the gate can sink sufficient current, this allows high-voltage low-current devices to be switched under logical control.

The 74HCXX series is another popular branch of the TTL family. It comprises CMOS versions of the 74XX devices. One of its main advantages is that, like CMOS logic (see next section), it operates on a wide range of voltages (2 V to 6 V) and does not require a regulated supply. It is fast-acting and has low power requirements.

CMOS LOGIC

This is based on complementary MOSFET transistors. By complementary, we mean that the transistors operate in pairs, one of the pair being n-channel and the other p-channel. Both are enhancement MOSFETs.

CMOS is available in two main series, in which the type numbers range from 4000 and 4500 upwards. The most commonly used series, the 4000B series, has buffered outputs. The 4000UB series has unbuffered outputs. Although the inclusion of buffers adds to the time that the gate takes to operate, the buffers make the output more symmetrical, with equal and faster rise and fall times. This is important when CMOS gates are used in pulse circuits and timing circuits.

TTL AND CMOS COMPARED

For all except the fastest logic, the choice of family usually lies between TTL and CMOS. Here we compare the characteristics of the most popular TTL series, the low power Schottky series, and the CMOS 4000B series. As already mentioned, the TTL 74HCXX series combines some of the advantages of both. Data sheets list many characteristics but the ones of most interest to the designer are the following:

- Operating voltage.
- Power consumed, usually specified as the power per gate.
- Input voltage levels, the level above which the gate recognises the input as logic high, and the level below which it is taken as logic low.

- Output voltage levels, the minimum high level, the typical high level, the maximum low level and the typical low level produced by the gate.
- Input currents, the maximum when a gate is receiving a high input and the maximum when it is receiving a low input.
- Output currents, the minimum when a gate has a high output and the minimum when the gate has a low output.
- Fanout, the number of inputs that can be driven from a single output (can be calculated from input and output currents).
- Propagation delay, the time, usually in nanoseconds, between the change of input voltages and the corresponding change in output voltage.
- Speed, as indicated by the maximum clock rate it is possible to generate.

The 74LSXX series, the 74HCXX series and the CMOS 4000B series have the characteristics listed in the table below. Their input and output voltages are shown in Figure 18.10.

The 74LS series operates on 5 V DC, which must be held to within 5%. This is the voltage commonly used in computers and similar equipment.

By contrast, CMOS is able to work on a wide voltage range, which makes it much more suitable for battery-powered equipment. It is also easier to interface with logic circuits and with other devices.

Even though 74LS is a low-power series compared with standard TTL (which uses 10 mW per gate), 74LS series needs much more power than CMOS, and the 74HC series. Again this gives CMOS and 74HC the advantage for battery-powering and for portable equipment in general.

Note that although the *average* power requirement of CMOS is low, the amount of power used depends on the amount of activity in the circuit. If the circuit is quiescent, with no logical changes occurring, it uses almost no power. But when it is changing state at high frequency the amount of power used may rise almost to the level used by TTL.

Characteristic	74LS00	74HC00	CMOS
Operating voltage	5 V ± 0.25 V	2 V to 6 V	3 V to 15 V (max 18 V)
Power per gate	2 mW		0.6 μW
Input current, high (max)	20 μA	0.1 μA	0.3 μA
Input current, low (max)	−0.4 mA	−0.1 μA	−0.3 μA
Output current, high (min)	−0.4 mA	−4 mA	−0.16 mA (−1.2 mA)
Output current, low (min)	8 mA	4 mA	0.44 mA (3 mA)
Fanout	20	50 (nominal)	50 (nominal)
Propagation delay	9 ns	10 ns	125 ns (40 ns)
Fastest clock rate	40 MHz	40 MHz	5 MHz

74LS   CMOS 4000
5V   5V   15V

4.95   14.95

3.5   11.0

2.7

2.0   1.5

0.8   4.0
       0.5

              0.05   0.05

I/P O/P   I/P O/P   I/P O/P

FIGURE 18.10   The shaded areas indicate the voltage ranges that are defined as logic low and logic high for TTL and for CMOS with a 5 V or a 15 V supply voltage. The output voltage levels of gates are set to be well within the limits accepted at the inputs. Levels for 74HC00 at 5 V are the same as for CMOS.

CMOS has insulated input gates (using 'gate' in the other sense) like any other FET device, so it is not surprising to find that input currents are very low in CMOS and 74HC. The output currents are lower than those of 74LS, the figures being quoted for a 5 V supply and (in brackets for CMOS) for a 15 V supply.

When calculating fanout we have to consider what happens under both high and low logic conditions.

Example

Consider a pair of 74LS gates, gate A which is feeding its output to an input of a second 74LS gate, gate B. If A is high, then, according to the table, it can supply at least 0.4 mA, or 400 µA. Only 20 µA are needed to supply the input of B, so A can supply 400/20 = 20 gates.

Now suppose that the output of A is low. It can draw at least 8 mA from any outputs to which it is connected. But gate B needs only 0.4 mA drawn from it to qualify as a low input. So the number of gates that A can bring down to a low level is 8/0.4 = 20. At both high and low levels A can drive the inputs of 20 gates. Its fanout is 20.

This is far more inputs than a designer would normally need to connect to a single output so, for practical purposes, the fanout is unlimited.

Self Test

What is the fanout when:
(a) a 74LS gate drives a 4000B series gate?
(b) a 4000B gate drives a 74LS gate?

Similar calculations can be done for CMOS but here we have the extra complication that output currents depend on operating voltage. The high input resistance of CMOS gates means that the fanout is large. A fanout of 50 is taken as a rule-of-thumb number, and is one that will very rarely be needed.

Looking at Figure 18.10, we note a major difference between 74LS and CMOS. On the left we see that a 74LS gate takes any input voltage (I/P) greater than 2.0 V to be logic high. Any input lower than 0.8 V is low. Inputs between 0.8 V and 2.0 V are not acceptable and give unpredictable results. Acceptable inputs are provided by 74LS outputs as shown in the O/P column.

A 74LS gate is guaranteed to produce an output of 2.7 V or more when it goes high. Thus any high output will register as a high input, with a margin of 0.7 V to spare. This allows for spikes and other noise in the system and gives a certain degree of **noise immunity**. Provided that a noise spike takes a high output level down by no more than 0.7 V, it will still be taken as a high level. In this case the noise immunity is 0.7 V. Noise immunity is very important in logic systems. A '0' that accidentally becomes interpreted as a '1' can have devastating effects on the logic.

At low level the maximum output voltage is 0.5 V, though typically only 0.25 V. This is 0.3 V (typically 0.55 V) below the input level that is registered as logic low, so again there is a useful amount of noise immunity.

In the case of CMOS, the high and low input levels vary with operating voltage. But the output levels are always within 0.05 V of the positive or 0 V supply lines. This is because the family is built from MOSFETs. When turned on, they act as a very low resistance between the positive supply line and the output. When turned off, they act as a very low resistance between the negative line and the output. This gives CMOS very high noise immunity.

The table shows that 74LS is much faster than CMOS and this allows clocks and other time-dependent circuits to run at much higher rates with 74LS. Propagation delay is reduced in CMOS if the supply voltage is made higher. The values quoted are for the more frequently used 4000B series. Part of the delay is due to the output buffers. If speed is important, the 4000UB series has a propagation delay of only 90 ns when used at 5 V.

EMITTER COUPLED LOGIC

The prime advantage of this family is its high speed, which is due to the fact that the transistors are never driven into saturation. This makes it the fastest of all logic families with a propagation delay of 1 ns per gate and a maximum clock rate of 500 MHz.

As well as avoiding transistor saturation, the circuits have low input impedances to avoid the speed-reducing effects of capacitance. Because of these measures, power consumption is high, as much as 30 mW per gate. Also, the logic output levels are close together at -0.8 V and -1.6 V. This leads to low noise immunity. Consequently, an ECL circuit needs an ample power supply carefully protected against voltage spikes. Another factor which makes ECL difficult to work with is that the fast edges on the signals mean that connecting wires must be treated as transmission lines, with precise routing on the circuit board.

Because of the difficulties associated with ECL, its use is restricted to large computer systems in which its high speed is an indispensable benefit.

ACTIVITY — LOGIC LEVELS

Investigate the input and output logic levels of logic gates. Use CMOS 4011, and 74LS00 dual-input NAND gates. You need:

- A breadboard.
- A supply of short lengths of single-core wire with insulation stripped from both ends.
- A power supply, 5 V DC for 74LS series IC, or any DC voltage between 3 V and 15 V for CMOS.
- Integrated circuits to provide the gates.
- Data sheets to show the pin connections of the ics.
- Output indicator, either a testmeter or a logic probe. You can also use an LED in series with a resistor, as on p. 149.

If NAND or NOR gates are used with inputs wired together they act as INVERT gates. Connect the two inputs of a gate together and supply them with input from the wiper of a variable resistor connected across the supply lines as a potential divider. The CMOS IC can be operated on any supply voltage in the range $+3$ V to $+15$ V.

Unused CMOS inputs must be connected to the supply or 0 V before testing. Unused TTL inputs should be connected to the supply through a 1 kΩ resistor; several inputs can share a single resistor.

Increase the voltage slowly from 0 V up to the supply voltage. Measure the input voltage at various stages and note whether the output is low or high. Record the level at which the output becomes firmly low. Decrease the voltage slowly from supply voltage to 0 V. Record the level at which the output becomes firmly high.

For the CMOS gate, repeat this investigation at two or three different supply voltages. Record your results in a table to compare the two families.

ACTIVITY — TRUTH TABLES

Investigate the behaviour of the basic logic gates, using CMOS, 74HC or 74LS gates. For each gate, draw up a blank truth table and mark in all possible combinations of inputs, using 1 and 0 or H and L to indicate the two possible logic states. A suitable test circuit is illustrated by Figure 18.11. Then test the combinations of gates shown below and identify the function that the combination performs.

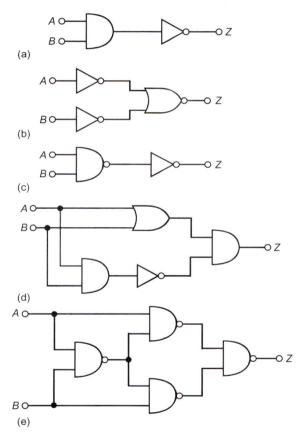

FIGURE 18.11 Test these combinations of gates as described in the Activity above.

On the Web

The TTL and CMOS families of logic circuits have grown with the development of new manufacturing technologies. There are over 25 subfamilies of CMOS ics, for example. Some are designed to operate on low voltage,

FIGURE 19.7   A run on a simulator confirms that the truth table for Figure 19.2 is correct. The display shows logic circuits (left) that produce the eight different combinations of inputs to the four logic gates (centre). The graphs show the sequence of logical input and output levels.

The trace for output Z shows an unexpected effect. There is a **glitch** in the output, shown as a brief low pulse occurring as input changes from 011 to 100. The reason for this is discussed in Topic 22. It is not a fault in the logic of the circuit.

WORKING WITH NAND AND NOR

Usually we are given a circuit or a truth table and we design a circuit to perform the logic with the minimum of gates. The circuit in Figure 19.3 is an example of a minimised circuit. Although it requires the minimum of components, takes up the least possible board space and uses the minimum of power, it may not be ideal in practice. When a circuit is built from ICs there is the practical point that an IC usually contains several identical gates. A single IC may contain six NOT gates, or four 2-input gates performing the other basic logical operations.

It is uneconomical of cost, board space and power to install a 6-gate or 4-gate IC on the circuit board if only one or two of its gates is going to be used. It is better to try to restrict the design to a few different types of gate and so reduce the number of ICs required.

This circuit requires only two gates, NOT and AND but unfortunately they are of different kinds. To set up this circuit on its own requires two ICs containing 10 gates, of which we use only 2. If this circuit is part of a larger logic system, we may find that the system does not need any other AND gates. Or, if it does need a few AND gates, they are situated a long way from the NOT gates. This means having long tracks wandering all over the board. Tracks take up space and, if they are long, may delay the signal sufficiently to cause the logic to fail. The more long tracks there are, the more difficult it becomes to design the board, and the more we need to resort to wire links and vias to make the connections.

Via

On a printed circuit board a via is a connection (usually a double-headed pin soldered at both ends) that goes through the board, linking a track on one side of the board to another track on the other side.

Circuit board design can be simplified by keeping to two or three types of gate. Better still, use only one type. All logic functions can be obtained with NOT, AND and OR, but these are not the most useful gates in practice. NAND and NOR gates too can be used to perform all the basic logic functions and are also useful in building flip-flops, latches and other more complicated logical circuits. For this reason, we generally prefer to work only with NAND and NOR gates. There are some circuits, such as decoders, that need a lot of AND gates, which makes it worth while to use ICs of this type. Some kinds of circuit need a lot of NOT gates. But, on the whole, NAND and NOR gates are preferred.

Example

The first step in converting the circuit of Figure 19.3 to NAND and NOR is to note that the second half of the truth table repeats the first half, so it is unnecessary to take C into account. This leaves two inputs, A and B, so there are four possible combinations of input state. Now work backwards from the output. The outputs for the four combinations comprise one 0 and three 1's. The truth tables on the first page of this topic show that this mix of outputs is a feature of NAND gates. Therefore the output gate of the converted circuit must be NAND.

Compare the normal output of NAND with the output of the circuit in Figure 19.3:

Inputs		Outputs	
B	A	NAND	Circuit
0	0	1	1
0	1	1	1
1	0	1	0
1	1	0	1

Inverting A in the bottom two lines makes the NAND output agree with the required output. Inverting A in the top two lines makes no difference as both lines have output 1. So the solution is to invert A before feeding it to the NAND gate. We can use a NOT gate if there is one to spare. Otherwise, build the circuit from two NAND gates, as below:

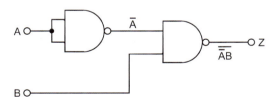

FIGURE 19.8 It is often useful to be able to work only with NAND gates. This is the NAND gate equivalent of the circuit in Figure 19.2 and its simplified version in Figure 19.3.

We can also use De Morgan's Theorem to help design a circuit in NAND or NOR gates. Figure 19.9 shows equivalent circuits.

Exclusive-OR and exclusive-NOR ICs are available, usually four gates to a package, and sometimes a circuit needs several such ICs (see the parity tree in

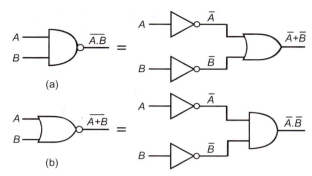

FIGURE 19.9 Replacing (a) OR with NAND, and (b) AND with NOR, after inverting the inputs.

Topic 20). But often we need such a gate only once or twice in a circuit and it is better then to build it from other basic gates. The Karnaugh map technique is no help as there are no entries from the truth tables that can be grouped together.

The expression for exclusive-OR from the truth table itself is:

$$A \oplus B = A\overline{B} + \overline{A}B$$

This can be translated directly into gates, using three of the basic types, NOT, AND and OR:

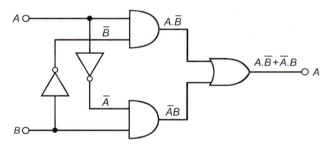

FIGURE 19.10 An exclusive-OR gate that is built from NOT, OR and AND gates, requires three different ICs.

In practice we might not find all three types already available as spare gates, and they might be too far apart on the circuit board to make it convenient to connect them.

An alternative exclusive-OR circuit circuit can be built entirely from NAND gates:

FIGURE 19.11 This version of the exclusive-OR gate uses only NAND gates to build it. It is assembled from a single IC.

Logical Combinations

Using the design techniques described in Topic 19, it is possible to build a wide range of logical functions. In this topic we look at a number of useful functions that you might need when designing electronic projects. All of these depend on logical combinations, often referred to as **combinatorial logic**. The output of a combinatorial logic circuit depends solely on the present state of its inputs. Its output can be predicted from its truth table.

Combinatorial logic contrasts with **sequential logic** which is fully explained in Topics 21 and 22.

Many of the functions you may need in project design can be achieved with fewer than ten gates. These you can build from the basic gates in IC packages. You might build them entirely from NAND gates.

Other functions might require dozens, or possibly hundreds, of logic gates. Designing, building and testing such circuits are difficult and demanding tasks.

Fortunately, the more complex logical functions are so frequently used that manufacturers have designed ICs in which complete logical systems are ready-built on a single chip. This is known as **medium scale integration** (MSI), in contrast to the **small scale integration** (SSI) of the NAND gates and the other simple logical devices that we met in Topic 19. In this topic we look at some of the more commonly used MSI devices for dealing with logical combinations.

MAJORITY LOGIC

It is sometimes useful to have a circuit with an odd number of inputs and for the output to take the state of the majority of the inputs. For example, with a 5-input majority circuit, the output is high if any 3 or 4 or all 5 inputs are high.

Inputs			Output
C	B	A	Z
0	0	0	0
0	0	1	0
0	1	0	0
0	1	1	1
1	0	0	0
1	0	1	1
1	1	0	1
1	1	1	1

The truth table of a 3-input majority circuit (above) requires three inputs, A, B, and C. The output is high when any two of A, B, and C are high or when all three are high.

Stating this as a logical equation, we obtain:

$$Z = \overline{A}BC + A\overline{B}C + AB\overline{C} + ABC$$

To implement this as written would take three NOT gates, four 3-input AND gates and a four-input OR gate, a total of 8 gates. Here is the Karnaugh map of the circuit:

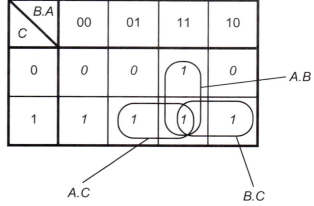

FIGURE 20.1

Another advantage of this technique is that the output for any particular combination of inputs can easily be altered, simply by connecting the corresponding input to the other supply rail. If we were using an array of gates, a change of logic is likely to require gates of a different type, and re-routed connections.

The 4512 data selector has **tri-state outputs**. When the inhibit (INH) input is low, the output pins behave in the normal way, outputting either high or low logical levels. When the inhibit input is made high, *all* the outputs go into the **high impedance** state. In this state, a high impedance exists between the output of the gate and the external line to which it is normally connected. In effect, all outputs become disconnected from the system and no signals come from the device.

This feature is in common use in computers and other data processing systems, where several ICs may be wired to a **data bus**. This bus carries data to many other parts of the system. But other parts of the system must receive data from only one device at a time. All the devices that put data on to the bus have tri-state outputs, but only the one device that is supposed to be transmitting data at that time has its inhibit input made low. All the others have their inhibit inputs high and are, in effect, disconnected from the bus.

> **Data Bus**
>
> A set of conductors between any two or more parts of a data-processing system, often each carrying the individual bits of a data group.

USING ROM

Another approach to complicated combinational logic is to use a ROM. A memory chip has address inputs by which any item of data stored in the memory can be read. Chips vary in the number of data items stored and the number of bits in each item.

> **ROM**
>
> A read-only memory device.

An example of the smaller ROMs is the 82S123A, which stores 256 bits arranged in 32 bytes, each consisting of 8 bits. Any particular byte is read by inputting its address (00000 to 11111, equivalent to 0 to 31 in decimal) on five address input pins.

When the 'chip enable' input is made low, the byte stored at this address appears on the eight tri-state output pins. Thus, for each of the 32 possible combinations of input states, we can obtain one of 32 corresponding output bytes. There is no need for complicated logic circuits to decode the inputs. We simply program the ROM with the required output levels.

A device such as the 82S123A can be used to control up to eight different devices such as lamps, displays, solenoids, or motors. It is programmable only once, but many other types of ROM may be reprogrammed if the logic has to be changed.

PROGRAMMABLE ARRAY LOGIC

If a logic design is highly complicated it may be best to create it in a **logic array**. An array has several hundreds or even several thousands of gates, all on one chip. They may be connected together by programming the device, using a personal computer. The chip is usually contained in a PLCC package, which is square in shape with a large number (as many as 84) pins arranged in a double row around the edge of the device. Depending on the type of PAL, some of the pins are designated as inputs, some as outputs, and some may be programmable to be normal inputs or outputs or to be tri-state outputs. Many PALs also contain an assortment of ready-made flip-flops and other useful sequential logic units.

Programming PALs is an intricate matter. It is done on a development board or programmer, which provides the programming signals. This is under the control of a personal computer, running special software to produce the required logical functions. Most types of PAL are erasable and reprogrammable.

LOGIC IN SOFTWARE

Mention of software reminds us that there is an increasing tendency to use microprocessors and microcontrollers (such as the PIC and the AVR controllers) to replace complicated logic circuits. We use *software* logic instead of *hardware* logic. This allows even more complicated logical operations to be performed, yet requires no greater expense, board space or power.

As explained in detail in Topic 36, the microcontroller receives inputs from sensor circuits and control switches. The logic is performed by software running on the microcontroller. The truth table is built into the logic of the software. This determines what the outputs will be. Then the appropriate signals are sent to the microcontroller's output terminals. If the logic needs to be changed or updated, it is relatively simpler and quicker to re-write software than it is to redesign and reconstruct hardware.

ACTIVITY — MSI DEVICES

Investigate some of the MSI devices that perform the combinational functions described in this topic. Work on a breadboard or use a simulator.

Devices to investigate include:

Adder (4-bit, full): 4008, 74LS283.
Magnitude comparator: 4063 (4-bit), 74LS85 (4-bit), 4512 (8-bit).
Data selector/multiplexer: 4019 (2 line), 74LS157 (2-line), 74LS153 (4-line), 74LS151 (8-line), 4512 (8-line).
Data distributor/demultiplexer/decoder: 74LS138 (3-to-8 line), 74LS139 (2-to-4 line), 4555 (2-to-4 line).
Decoder: 4028 (1 of 10), 4511 (7-segment), 74LS47 (7-segment).
Priority encoder: 4532 (8-to-3 line), 74LS148 (8-to-3 line), 74LS147 (10-to-4 line).

Verify that the circuit operates according to the truth table (function tables) in the manufacturer's data sheet.

QUESTIONS ON LOGICAL COMBINATIONS

1 Design a 3-bit binary to decimal converter, using only NOR gates.
2 Explain the action of a 2-input data selector.
3 Design a 3-input data selector, with an enable input and TRUE and NOT outputs.
4 Design a 4-input data selector, using only NAND gates.
5 Design and verify (on a breadboard or simulator) a circuit for driving a 7-segment display for (a) the digits 0 to 7, and (b) the even digits 0 to 8.
6 Explain how a binary-to-decimal decoder can be used as a data distributor.
7 In a shop, a siren is to be sounded automatically if:
  - the emergency button is pressed, or
  - smoke is detected, or
  - a person is detected inside the shop between dusk and dawn, or
  - if someone enters the storeroom without first closing the hidden 'storeroom access' switch, or
  - the shopkeeper leaves the premises without first locking the storeroom door.

Design the logic for this system. State what outputs each sensor and switch gives to indicate the 'alarm' condition.

8 In a greenhouse, the irrigation system is to be turned on automatically:
  - when the temperature rises above 25°C, or
  - when the humidity falls below 70%, or
  - between 14:30 hrs and 15:30 hrs every day.
  The system must never be turned on when:
  - it is between dusk and dawn, or
  - the temperature falls below 7°C,
  - there is someone in the greenhouse, or
  - the manual override switch has been closed.
  Design the logic for this system. State what outputs each sensor and switch gives to indicate the 'irrigate enable' condition.

9 Outline the techniques available for building circuits for arbitrary truth tables when the logic is complicated.

MULTIPLE CHOICE QUESTIONS

1 The output of a majority logic circuit goes high when:
  A more than half the inputs are 1.
  B all inputs are 0.
  C some inputs are 1 and some are 0.
  D all inputs are 1.
2 The inputs to a half adder are $A = 1$ and $B = 1$. Its outputs are:
  A $S = 1$, $C_o = 1$.
  B $S = 0$, $C_o = 1$.
  C $S = 1$, $C_o = 0$.
  D $S = 0$, $C_o = 0$.
3 The outputs of a full adder are $S = 0$, $C_o = 1$. Its inputs might be:
  A $A = 1$, $B = 0$, $C_i = 0$.
  B $A = 1$, $B = 1$, $C_i = 1$.
  C $A = 1$, $B = 0$, $C_i = 1$.
  D $A = 0$, $B = 0$, $C_i = 0$.
4 An 8-input parity tree has odd parity. Its output goes high when the input is:
  A 11111111.
  B 10011000.
  C 10000001.
  D 00000000.
5 'Demultiplexer' is another name for:
  A a parity tree.
  B a data selector.
  C a decoder.
  D a data distributor.

has a 11 input, and its output is about to change to 0. The flip-flop can not stay in this state; it is unstable.

(c) The output of Gate A has gone low, so that the input to Gate B is now 00. This makes no difference to its output, which stays high. The circuit is now in the *Reset* state, with Q = 0 and $\overline{Q}$ = 1. It is *stable* in this state because both gates have inputs and outputs that conform to the NAND truth table.

(d) There is no further change of output when the Reset input goes high again. This is because the output of B remains high when it has 01 input. The flip-flop remains stable.

FIGURE 21.3   This edge-triggered circuit produces a low output pulse when the input rises from low to high. Its output remains high when input falls from high to low.

Summing up, $\overline{Q}$ is the inverse of Q in the stable states. The flip-flop is made to change state by applying a low pulse to its Set or Reset input, but it does not change state when an input receives two or more low pulses in succession.

A similar flip-flop is built from a pair of NOR gates. Its inputs are normally held low and it is made to change state by a high pulse to its Set or Reset inputs.

A flip-flop stays in the same state until it receives an input that changes its state. It can 'remember' which kind of input (Set or Reset) it received most recently. Flip-flops are the basic circuit used for registering data, particularly in computer memory circuits.

DELAYS

Another way in which the behaviour of a logic circuit can be made to go through a sequence of changes is by introducing a **delay**. Usually this is done by including a capacitor, which takes time to charge or discharge.

Delay circuits have already been described in Topic 5, but here we shows an example based on logic gates. It is built from two 2-input NAND gates. Gate A has its inputs connected together so that it behaves like a NOT gate.

FIGURE 21.4   The sequence of stages in the generation of a pulse by the delay circuit. The letters (a)–(d) indicate the stages in the timing diagram.

The way the delay works is shown in Figure 21.4 and in the timing graphs in Figure 21.5. The sequence of stages is as follows:

(a) The normal input to the circuit (dark grey line) is low (0 V). This gives Gate A a high output (15 V). The capacitor has charged to 15 V through the resistor. So the input to Gate B is 01, giving it a high output (black line). This is a stable state.

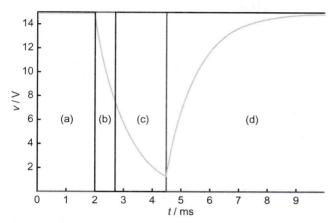

FIGURE 21.5 Changes of voltage in the delay circuit show a rise of input at 2 ms (light grey), which immediately produces a low output pulse lasting 0.7 ms (black). Output remains high when input goes low at 4.5 ms. Changes of voltage across the capacitor are plotted in dark grey.

(b) At 2 ms the input to the circuit (dark grey line) goes high. This immediately changes the output of Gate A from 1 to 0. The capacitor begins to discharge through R1, current flowing *into* the output of Gate A. The exponential fall in voltage across the capacitor can be seen in the timing diagram (light grey line). Nothing happens to Gate B (black line) at this stage.

(c) At 0.27 ms the falling voltage across the capacitor reaches 7.5 V, so the input to Gate B is now equivalent to 01. Its output goes high, ending the low output pulse.

(d) When the input pulse ends (here at 4.5 ms, but it could be at *any* time after the beginning of stage (c)) the state of Gate A is reversed and the capacitor begins to charge again. It makes no difference to the output of Gate B when its input changes from 00 to 01. So there is no output pulse when the input pulse ends.

This circuit is often used for generating pulses in logic circuits. The length of the pulse depends on the values of the resistor and capacitor.

Self Test

Describe the action of a pulse generator based on a pair of NOR gates.

The chief disadvantages of this simple monostable are:

- It must be given time to completely recharge the capacitor before being triggered again.
- The input must go high for longer than the length of the output pulse.
- Pulse length depends on supply voltage.

The delay circuit is stable in its quiescent or 'ready' state, so it is a monostable (p. 41). Once triggered, it produces a pulse and then returns to the same stable state. Monostables with higher precision are available in the 74LS and CMOS series.

MONOSTABLES

The 74LS221 (below) is a typical monostable pulse generator. The pulse length is determined by the values of the resistor and capacitor and is approximately $0.7RC$ seconds. This is the time taken to charge the capacitor. Using a capacitor of maximum value 1000 µF and a resistor of maximum value 70 kΩ, the greatest pulse length obtainable is 70 s. Output Q is normally low and $\overline{Q}$ is normally high. The monostable is triggered by a falling edge on input A or a rising edge on input B, as indicated by the truth table on the next page.

FIGURE 21.6 The 74LS221 comprises two identical monostables. The pin numbers in this figure refer to only one of these.

This is the truth table for the operation of the 74LS221 monostable IC:

Inputs			Outputs	
CLR	A	B	Q	Q̄
0	X	X	0	1
X	1	X	0	1
X	X	0	0	1
1	0	↑	⊓	⊔
1	↓	1	⊓	⊔
↑	0	1	⊓	⊔

The CLEAR input is active low. When it is made low, the values of A and B have no effect; output Q is low and output Q̄ is high. The monostable can also be cleared by making A high or B low, whatever the state of the CLEAR input.

To generate a pulse the CLEAR input must first be made high. Then a falling edge on A (a downward arrow in the table) or a rising edge on B (a rising arrow) triggers the pulse to begin. Once the monostable is triggered, the length of the pulse depends only on the values of R and C. In other words, the monostable is not retriggerable. However, a low level on CLEAR can end the pulse at any time. The last line of the table shows that a pulse can also be triggered by the CLEAR input. This only occurs if certain inputs have already been made to A and B and is not part of the normal action of the circuit.

The 74LS221 has features that make it superior to the circuit on p. 173:

- It requires very little time to recover at the end of the output pulse. Under the right conditions it can have a duty cycle up to 90%. That is, it recovers in about one-ninth of the length of its output pulse.
- B is a Schmitt trigger input. This means that the monostable is reliably triggered on a slowly changing input voltage level.
- The length of the timing period does not depend on the supply voltage.
- The length of the timing period does not depend on temperature.

The CMOS 4528 is a monostable with similar properties, except that it is re-triggerable, and its recovery time is usually several milliseconds.

The 555 timer IC is generally the most popular choice for monostables and for astables, as described in Topic 5.

CMOS ASTABLES

Astables can be built from TTL and CMOS gates. The diagram below shows one built from a pair of CMOS NOT gates. The third gate is not part of the astable but acts as a buffer between the astable and an external circuit such as an LED. R2 is not an essential part of the astable, but serves to give the output signal a squarer shape. It has a value about 10 times that of R1.

FIGURE 21.7 A pair of NOT gates (or their equivalent made from NAND or NOR gates) make an astable multivibrator. The third gate (grey) acts as a buffer.

The action of the astable is as follows, referring to the diagram below. (To avoid having to repeat the words 'the supply voltage' several times, assume that it is 15 V):

(a) The input of Gate 1 is 0 V, its output is 15 V, and therefore the output of Gate 2 is 0 V. The capacitor is at a low voltage (actually negative, but see later). Current flows from Gate 1 through R1 and gradually charges the capacitor. The rate of charging depends on the values of R1 and C1.

FIGURE 21.8 The inverting action of the NOT gates is used to charge and discharge the capacitor. The time taken to do this, and therefore the frequency of the astable, depends on the values of the capacitor and resistor R1.

(b) When the voltage across the capacitor reaches about 7.5 V, the input to Gate 1 (which is at almost the same voltage) is taken to be at logic high. Instantly, the levels change to those shown in (b). The sudden rise in the output of Gate 2 pulls *both* plates of the capacitor up by 15 V. This means that the plate connected to R1 rises to 22.5 V. Now current flows in the reverse direction through R1 and *into* the output of Gate 1. The voltage at this plate falls until it has reached 7.5 V again. The input of gate 1 is taken to be logic low and all the logic levels change back to state (a) again. The sudden fall in the output of Gate 2 pulls both plates of the capacitor down by 15 V. The plate connected to R1 falls to −7.5 V. The astable has now returned to its original state, stage (a).

The frequency of this astable is given by:

$$f = 1/2.2R_1C_1$$

BISTABLE LATCH

A latch is a circuit is which the output is normally equal to the data input but which can be held (latched) at any time.

In the diagram below the gates on the right form an S-R flip-flop exactly as Fig 21.2.

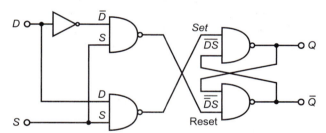

FIGURE 21.9 With a data latch, the Q output follows the data input (D) when the STORE input (S) is high, or is latched when S is low.

In that circuit the inputs are normally held high and a low pulse to the Set input sets the flip-flop. In this circuit the Set input of the flip-flop receives $\overline{D \bullet S}$ while the Reset input receives $\overline{\overline{D} \bullet S}$. D is the Data input which at any instant is 0 or 1. S is the Store input. The truth table for D and S is:

Inputs		Outputs		Effect on Q
S	D	$\overline{D \bullet S}$	$\overline{\overline{D} \bullet S}$	
0	0	1	1	Latched
0	1	1	1	Latched
1	0	1	0	Equals D
1	1	0	1	Equals D

When $S = 0$ the outputs of the NAND gates on the left are 1. They hold the flip-flop in the latched state, since neither input can go low. When $S = 1$, and D goes low, the Reset input goes low, resetting the flip-flop so that Q goes low. When $S = 1$ and D goes high, the Set input goes low, setting the flip-flop so that Q goes high. Whatever the value of D, output Q is equal to D. The other output, Q, always is the inverse of Q and therefore of D.

There are ICs that contain several data latches equivalent to the above. They have individual data inputs, Q outputs and sometimes $\overline{Q}$ outputs as well. Examples are the CMOS 4042 and the 74LS75, which latch 4 bits of data simultaneously.

CLOCKED LOGIC

In the bistable circuits that we have looked at so far in this topic, the circuit changes state immediately there is a suitable change of input. This can cause difficulties in large logic circuits.

Logic gates always take time to operate and different amounts of delay may arise in different parts of a circuit owing to the different numbers of gates through which the signals have to pass. It becomes very difficult to make sure that all the bistables change state in the right order and at the right times to respond to the data.

The solution to this is a **system clock**. This controls the exact instant at which the bistables change state. It is like the conductor of an orchestra, keeping all the players in time with each other by beating 'time'. We say that the players (or the bistables) are **synchronised**.

Clocked logic devices usually operate on the **master-slave** principle. They are available as MSI integrated circuits. Each flip-flop actually consists of two flip-flops, the **master** and the **slave**. The master flip-flop receives data and is set or reset accordingly. The state of the master flip-flop can not be read on the output terminals of the IC at this stage. Then, on the next positive-going edge of the clock input, the data on the master is transferred to the slave and appears at the output terminals of the IC.

After this, new data may change the state of the master but this has no effect on the slave or on the data appearing at the terminals. The present output state is held unchanged until the next positive-going clock edge. Then the data currently on the master is transferred to the slave.

DATA-TYPE FLIP-FLOP

The symbol for the data-type (or D-type) flip-flop is shown in Figure 21.10 and its action is plotted in the graphs following. For a **clocked mode** operation such as this, the Set input (S) and Reset input (R) are held low. Some kinds of D-type flip-flop do not have the R input.

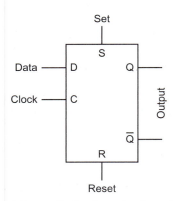

FIGURE 21.10   A D-type flip-flop has a data input (D), a clock input (C), and a Q output. It may also have set and reset inputs and an inverted Q output. Data is transferred to the Q output on the rising edge of the clock.

The graphs below are obtained from a simulator, modelling a 74LS74 D-type flip-flop. The clock is running at 10 MHz and the time scale is marked in nanoseconds. The lowest graph shows output Q. The simulator is stepped manually so that the data input can be made high or low at any desired time.

FIGURE 21.11   In this plot of the action of a D-type flip-flop, the top trace is the clock, with a period of 100 ns. The data (middle trace) can be changed at any time but the Q output (bottom trace) changes only on the rising edge of the clock. The change is slightly delayed because of the propagation delay of the flip-flop (22 ns in this case).

The run begins with both D and Q low. D is made high at 170 ms. At this time the clock is high. Nothing happens to the output when the clock goes low. But Q goes high at the next *rising edge* of the clock. This occurs at 250 ns. Looking more closely at the timing, we note that the rise in Q is slightly later than the rise in the clock. This is due to the propagation delay of the flip-flop, here set to a typical value of 22 ns. So Q eventually goes high at 272 ns. Subsequent changes in D, at 540 ns, 640 ns and 880 ns, always appear at output Q at the *next* rising edge of the clock (plus the propagation delay).

D-type flip-flops are used for sampling data at regular intervals of time. They then hold the data unchanged while it is processed. An example is the use of these flip-flops at the input of digital-to-analogue converters (see Topic 24).

The Set and Reset inputs are used in the **direct mode** of operation of the flip-flop. They are also known as the Preset and Clear inputs. When the set input is made high, Q goes high instantly, without waiting for the rising edge of the clock. Similarly, Q goes low immediately the Reset input is made high.

TOGGLE FLIP-FLOP

The output of a toggle (or T-type) flip-flop changes every time the clock goes high. Toggle flip-flops are not manufactured as such. It is easy to make one from a D-type flip-flop, by connecting the $\overline{Q}$ output to the D input.

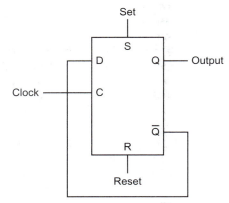

FIGURE 21.12   A D-type flip-flop with its $\overline{Q}$ output connected to its D input acts as a toggle flip-flop.

Assuming the D input at the most recent clocking was 0, then Q is 0 and $\overline{Q}$ is 1. So the D input is now 1. On the next clocking Q follows D and becomes high, so $\overline{Q}$ goes low. Now the D input is 0 and Q goes to 0 at the next clocking. The state of D and therefore of Q changes every time the flip-flop is clocked. Below is a graph of the toggle action.

FIGURE 21.13   The clock (top trace) is running at 10 MHz. The Q output of the toggle flip-flop changes state slightly after every rising edge of the clock. For this reason, the frequency of the output from the flip-flop is 5 MHz.

The toggle flip-flop demonstrates one of the advantages of clocked logic. After the flip-flop has been clocked, there is plenty of time before the next clocking for the circuit to *settle* into its stable state, and for its output to be *fed back* to the data input. This makes the operation of the circuit completely reliable. Unless the clock is very fast, there is no risk of its operation being upset by propagation time and other delays.

J-K FLIP-FLOP

The J-K flip-flop has its next output state decided by the state of its J and K inputs:

- If J = K = 0, the outputs do not change.
- If J = 1 and K = 0, the Q output becomes 1.
- If J = 0 and K = 1, the Q output becomes 0.
- If J = K = 1, the outputs change to the opposite state.

The $\overline{Q}$ output is always the inverse of the Q output. With most J-K flip-flops, the changes occur on the next rising edge of the clock. In the fourth case above, in which $J = K = 1$, we say that the outputs *toggle* on each rising edge.

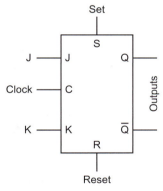

FIGURE 21.14 The output of a J-K flip-flop changes state on the rising edge of the clock (on the falling edge in a few types). Whether it changes or not depends on the inputs to the J and K terminals.

J-K flip-flops also have Set inputs and (usually) Reset inputs. These operate in the same way as those of the D-type flip-flop, producing an instant change of output, without waiting for the clock. Several types of J-K flip-flop are clocked by a negative-going edge instead of the more usual positive-going edge.

ACTIVITIES — FLIP-FLOPS

Investigate each of the circuits listed below, either on a breadboard or by using a simulator. Use the suggested ICs and consult data sheets for connection details.

Build a flip-flop using NAND gates as on p. 181 (CMOS 4011 or 74LS00).

Design and build a flip-flop using NOR gates (CMOS 4001 or 74LS02).

Build a pulse generator using NAND gates as on p. 182.

Design and build a pulse generator built as on p. 182 but with NOR gates.

Design and build a pulse generator which produces a low pulse on a *falling* edge, using two NAND gates.

Investigate a pulse generator built from the 74LS221, with pulse length 1.5 ms. Check the pulse length with an oscilloscope.

Design and build a CMOS astable as on p. 183 using either NAND (4011), NOR (4001) or NOT (4069) gates. Select components to give a frequency of 100 Hz and check the frequency by using an oscilloscope.

Build a CMOS astable using exclusive-OR gates (4070). Construct the bistable latch shown in Figure 21.9 and study its action by placing probes at various points in the circuit.

Investigate an MSI data latch (CMOS 4042, 74LS75.

ACTIVITIES — CLOCKED LOGIC

Investigate each of the circuits listed below, either on a breadboard or by using a simulator, using the suggested ICs. Consult data sheets for connection details. Most of the ICs specified below contain two or more identical flip-flops, of which you will use only one. With CMOS, it is essential for *all* unused inputs to be connected to the positive or 0 V rail.

Check out a D-type flip-flop (CMOS 4013, 74HC74, 74LS174, or 74LS175). Use a slow clock rate, say 0.2 Hz. Connect the data input to the positive rails through a 1 kΩ resistor to pull it high. Use a push-button connected between the data input and the 0 V rail to produce a low input. The 74LS174 has only Q outputs but the others have Q and $\overline{Q}$ outputs. Investigate both clocked and direct modes.

> **Flip-Flops**
>
> Animated diagrams of D-type and JK-type flip-flops are displayed on the Companion Site.
> There is also a diagram of a CMOS astable.

Test a J-K flip-flop (CMOS 4027, 74LS73A, 74LS112A, 74LS379). Connect J and K inputs with pull-up resistors and push-buttons.

Make a toggle flip-flop from a D-type flip-flop or a J-K flip-flop, and investigate its action.

QUESTIONS ON LOGICAL SEQUENCES

1 Explain the working of a flip-flop built from two NAND gates.
2 Explain how to make a pulse generator based on two NAND gates.
3 Describe how to make an astable using two CMOS NOT gates, a capacitor and two resistors.
4 Show how to build a data latch using logic gates and use a truth table to explain how it works.
5 What are the advantages of using clocked master-slave logic devices?
6 Draw a timing diagram of a toggle flip-flop, showing the input from the clock and the output from the flip-flop.
7 Calculate the value of the timing resistor required in a CMOS astable, given that it has a 4.7 nF capacitor and the frequency is 5 kHz. What is the nearest E24 value for the resistor? What is the E24 value of the other resistor (R2) in the astable?
8 Figure 21.7 shows NOT gates used in the astable. What other types of gate could be used, and how would they be wired?

MULTIPLE CHOICE QUESTIONS

1 A logical circuit with two stable states is NOT called a:
  A monostable.
  B flip-flop.
  C multivibrator.
  D bistable.
2 If the supply voltage is 12 V, the voltage across the capacitor of a 555 monostable as the pulse ends is:
  A 12 V.
  B 4 V.
  C 3 V.
  D 8 V.
3 If the supply voltage is reduced to 7.5 V in the monostable described in question 2, the length of the pulse is:
  A unchanged.
  B doubled.
  C halved.
  D divided by 3.
4 If the symbol for an input terminal has a bar drawn over it, it means that:
  A the input must always be low.
  B the input does nothing when it is high.
  C a negative input produces a negative output.
  D the input is made low to bring about the named response.
5 Which of these statements about a toggle flip-flop is NOT true?
  A The $\overline{Q}$ output is the inverse of the Q output.
  B The output follows the D input.
  C Clocking reverses the state of both outputs.
  D The $\overline{Q}$ output is connected to the D input.
6 A J-K flip-flop has its J and K inputs connected to the 0 V line. Its Q output is high. It is triggered by positive-going edges. When will its Q output go low?
  A on the next positive-going clock edge.
  B when the Set input is made high.
  C on the next negative-going clock edge.
  D never.
7 A circuit in which the output follows the input, but can be held when the store input goes low, is called:
  A a bistable latch.
  B an astable.
  C a toggle flip-flop.
  D a data flip-flop.
8 The $\overline{Q}$ output of a D-type flip-flop is made high by:
  A a low pulse on the D input.
  B a low-going edge on the clock input.
  C a high pulse on the reset input.
  D a low pulse on the preset input.

Counters and Registers

In Topic 21 we showed how a data-type (or D-type) flip-flop could be connected to make it into a toggle-type (or T-type) flip-flop. This is simply done by connecting its $\overline{Q}$ output to its D input. As a T-type flip-flop, its output changes state whenever its clock (or C) input changes from low to high — that is, it is triggered on a **rising edge**.

It is possible to connect two or more T-type flip-flops, feeding the Q output of one into the C input of the next. The drawing below shows three D-type flip-flops, operating as toggle flip-flops, and connected in a chain. We are basing the circuit on CMOS 4013 ICs, which contains two D-type flip-flops each, so this circuit needs two ICs, with one flip-flop to spare.

FIGURE 22.1 Three D-type flip-flops wired as toggle flip-flops act as a 3-bit up counter divider.

Assume that all the flip-flops are reset to begin with. This could be done by making their direct Reset inputs (not shown in the diagram) high for an instant. The flip-flops begin with their Q outputs low and their $\overline{Q}$ outputs high.

Clocking pulses, perhaps from an astable circuit, are fed to the clock input of the first flip-flop. Nothing happens until the clock changes from low to high. Then the flip-flop changes state. Its Q output goes high and its $\overline{Q}$ output goes low.

The change in its Q output from high to low has no effect on the second flip-flop, because it too is rising edge triggered.

There is no further change in the system until the next rising edge of the clock. Then the Q output of the first flip-flop changes back to low and Q output changes to high. This change from low to high triggers the second flip-flop to change state. The process continues, with the result shown in these graphs:

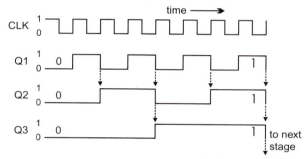

FIGURE 22.2 The graphs show the Q output from the flip-flops. The $\overline{Q}$ outputs are the inverse of these.

To sum up the action, we can say that:

- The outputs change state on the rising edge of the clock.
- Each Q output changes state on the falling edge of the previous stage (that is, on the falling edge of its $\overline{Q}$ output).

DIVIDING AND COUNTING

The action of the circuit can be looked at in two different ways:

- **Dividing:** each output changes state at half the rate of the clock or previous stage. The circuit acts as a **frequency divider**. For example, if the clock is running at 100 Hz, the frequencies of Q1, Q2 and Q3 are 50 Hz, 25 Hz and 12.5 Hz respectively.
- **Counting:** The Q outputs run through the sequence 000, 001, 010, 011, 100, 101, 110 and 111, repeating. These are the binary numbers from 0 to 7.

The wiring for all six AND gates is fairly complicated but, fortunately, a 10-stage walking ring complete with decoding is available as an IC, the 4017. The graph below shows its output sequence:

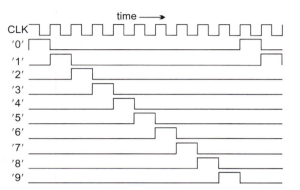

FIGURE 22.16 As the counter is incremented, the outputs of the 4017 go high one at a time, in sequence.

This type of counter is useful for triggering a sequence of events. As an illustration of the way in which it might be used, consider a fruit-canning machine. The process is divided into ten stages, repeating for each can processed. The stages might be those listed in the table below.

The clock drives the whole system. At each stage, a high pulse going to one of ten logic circuits triggers off the appropriate stage. At some stages the pulse may simply trigger a monostable to run a motor for a set length of time. At other stages it may trigger a more complex logical operation.

Stage	Operation
0	Put empty can on platform.
1	Fill with cooked cherries from a hopper.
2	Shake can to settle cherries.
3	Top up with more cherries.
4	Weigh can; reject if underweight.
5	Fill can with syrup.
6	Place lid on can.
7	Seal lid on to can.
8	Fix label on can.
9	Send can to packer.

This is an imaginary and simplified example, but it demonstrates the principles involved in using this counter. It is assumed that the frequency of the clock allows time for each operation to be completed. However, it would be possible to disable the clock automatically at any stage until that stage had been completed.

This IC has a 10-stage circuit but it can easily be made to operate on a shorter cycle. It has a reset input. This input is normally held low but, when it is made high, all flip-flops reset, so taking the device to stage 0. It continues counting from 0 when the reset input is made low again.

If an output from the counter is wired to the reset input, the counter resets each time that that stage is reached. For example, if output 8 is connected to the reset, the counter resets immediately output 8 goes high. Outputs 0 to 7 go high in turn, repeating. It acts as a modulo-7 counter.

The output graph in Figure 22.16 assumes that all Q outputs will be low when the circuit is first switched on. This does not necessarily happen. In the 6-stage walking ring illustrated in Figure 22.14 the circuit goes through six diffent states. These are 000, 100, 110, 111, 011, and 001.

With three flip-flops there are eight possible states. States 010 and 101 are not normally part of the sequence. These are the **disallowed states**.

On the first clocking, state 010 changes to 101 (shift first two digits to the right and invert the third to become the first). Conversely, state 101 changes to 010. If the counter is in one of the disallowed states at switch-on, it will alternate indefinitely between one and the other. The solution it to use two more AND gates to decode the disallowed states. Their outputs go to the reset input to clear the ring to state 000.

A walking ring can also be used to generate other output sequences, including arbitrary sequences and those with two or more outputs going high.

A Gray code generator is an example of a counter with arbitrary output. With a 4-bit binary counter the sequence is 0000, 0001, 0010, ... 1101, 1111. Sometimes two digits change at the same time, for example, when 0001 (count 1) changes to 0010 (count 2). Two bits change when 0011 (count 3) changes to 0100 (change 4). If the bits do not change at exactly the same time we will get intermediate stages, for example:

$$0011 \rightarrow 0001 \rightarrow 0101 \rightarrow 0100$$

Expressed as counts:

$$3 \rightarrow 1 \rightarrow 5 \rightarrow 4$$

The two transitory stages may last long enough to be decoded and so introduce glitches into the system.

In a Gray code, the outputs always change one at a time, which avoids the possibility of glitches. An example of a 3-bit Gray code sequence is:

$$000 \rightarrow 001 \rightarrow 011 \rightarrow 111 \rightarrow 101 \rightarrow 100$$

repeating 000...

Decoding the output from the ring is more complicated than for a 1-of-n counter, and is likely to need more gates. Decoding must also detect and correct for disallowed states.

STATIC RAM

For storing a large amount of data we use an array of many flip-flops on a single chip. A large array of this kind is usually called a **memory**. Each flip-flop stores a single bit of data. Each flip-flop can have data stored into it individually.

The flip-flop is either set or reset, depending on whether the bit is to be a 0 or a 1. This is called **writing,** because the process of storing data has the same purpose as writing information on a piece of paper.

Each flip-flop can also be **read** to find out whether the data stored there is 0 or 1. A RAM chip may have several hundreds or even thousands of flip-flops on it. To write to or read from any particular one of these, we need to know its **address**. This is a binary number describing its location within the array on the chip. This is discussed in more detail later.

Because at any time we can write to or read from *any flip-flop we choose*, we say this is **random access**. A memory based on this principle is a **random access memory**, or **RAM**. This is different from a SISO shift register, from which the bits of data must be read in the same order in which they were written.

Once data is stored in a flip-flop it remains there until either it is changed by writing new data in the same flip-flop, or the power supply is turned off and all data is lost. This type of RAM is known as **static RAM,** or **SRAM**.

DYNAMIC RAM

There is another type of RAM known as **dynamic RAM**, or **DRAM**. In this, the bits are not stored in flip-flops but as charges on the gate of MOSFET transistors. If a gate is charged, the transistor is on; if the gate has no charge, the transistor is off. Writing consists of charging or discharging the gate. Reading consists of finding out if the transistor is on or off by registering the output level (0 or 1) that it produces on the data line.

The storage unit of a DRAM takes up much less space than a flip-flop, so more data can be stored on a chip. Also it takes less time to write or read to a DRAM, which is important in fast computers.

The main problem with DRAMs is that the charge on the gate leaks away. The data stored there must be **refreshed** at regular intervals.

Under the control of a clock running at about 10 kHz, the charge is passed from one transistor to another, and the amount of charge is topped up to the proper level at each transfer. The data is kept 'on the move', giving this type of memory the description 'dynamic'. The need to refresh the memory regularly means that a certain amount of the computer's operating time must be set aside for this, and the RAM can not be used for reading and writing data during this time. The operating system is therefore more complicated than that needed for SRAM, which is available for use at any time.

FLASH MEMORY

A flash memory chip consists of arrays of MOS transistors each of which has two gates. The transistor that forms one of the gates is completely insulated so that any charge of electrons placed on it remains there indefinitely. The transistor remains on or off indefinitely. Its state is read by a technique that employs the other gate. The two possible states of each transistor are equivalent to the binary logic values '0' and '1', and in this way the chip stores arrays of binary data.

The advantages of flash memory are:

- It needs no power supply to keep the data in memory. This makes it ideal for portable equipment.
- It is removable. An owner of a digital camera can have several flash cards in use. Flash memory used in a computer can be removed and locked away for security.
- It is not damaged by mechanical shock, unlike hard disks made of glass.
- Writing and reading so it is ideal for memories holding large amounts of data. Flash memories holding hundreds of megabits are easily obtainable; chips storing 8 gigabits have been developed.

The disadvantages are:

- Individual bytes can not be erased; flash memory is erased in blocks.
- It eventually wears out; typically, a memory can be erased 10 000 times.
- It is more expensive per stored bit than a hard disk.

The advantages outweigh the disadvantages, so flash memory has become very popular.

Flash memory is used for storing data in a digital camera. A typical flash card stores 256 Mb, which is enough for hundreds of high-resolution digital colour photos. Later, after the data has been transferred to a computer, or has been printed out on to paper as a photo, the data can be erased and a new set of photographs recorded.

A memory stick is a small device generally with a built-in plug that fits into a socket on a CD or DVD player or on a computer. It may contain as much as 1 GB of flash memory. It is a convenient way of transporting data between computers as well as filing away text and images. The text and illustrations of this book need about 200 Mb for storage, so such a memory stick can store five books the size of this one.

Flash memory is also widely used in MP3 players, Personal Digital Assistants and mobile phones.

MEMORY CONTROL

Different types of RAM store different numbers of bits arranged in different ways. We take the 2114 RAM as an example. This is not *typical* since it is one of the smallest memory ICs commonly available. But it illustrates the principles of storing data.

FIGURE 22.17 The ten address inputs of the 2114 RAM allow for storage of 2^{10} (= 1024) words of data, each of 4 bits.

The 2114 stores 4096 bits (usually referred to as 4 kbits), which are writable and readable in groups of four. We say that the chip stores 1024 4-bit **words**. The chip has four data terminals (D0 to D3) through which a word may be written or read. Since there are 1024 words, each needs its own address. 1024 is 2^{10} so we need 10 address lines, A0 to A9. Then we can input every address from 0 (binary 0 000 000 000, all inputs low) to 1023 (binary 1 111 111 111, all inputs high).

As well as the connections to the data and address busses, there are two pins connected to the control bus. One of these is CHIP SELECT ($\overline{CS}$). The bar over its name and symbol indicate that this must be made low to have the desired effect. When it is high the chip is

unselected; its data terminals are in a high-impedance state so the chip is effectively disconnected from the data bus. Before we can write to or read from the RAM the $\overline{CS}$ input must be made low.

The other control input to the RAM is the WRITE ENABLE ($\overline{WE}$) input. The data terminals act either as inputs or outputs, depending on whether the $\overline{WE}$ pin is made high or low. With $\overline{CS}$ low and $\overline{WE}$ high, the chip is in **read mode**. Next we put a 10-bit address on the address bus. This selects one of the words and the 4-bits of data that it comprises are put on to the data bus.

With $\overline{CS}$ low and $\overline{WE}$ low, the chip is in **write mode**. Data present on the bus is stored in the addressed word in memory.

Often in computing circuits we handle data in bytes, consisting of 8-bits. Some RAMs are organised to store data in this format. For example, the 6264 RAM stores 64 kbits, organised as 8192 bytes, each of 8 bits. This needs 8 data input/outputs and 13 address lines.

In most SRAMs the bits are addressed individually, not as words or bytes. The 4164 SRAM for example stores 64 kbits as single bits. The chip needs only one data input/output, but 16 address lines.

On the whole SRAMs have shorter access times than DRAMs. The 2114 is relatively slow, taking 250 ns to write or read data. Most DRAMs are faster than this, averaging around 100 ns. By contrast, most SRAMs have access times of 60 ns.

ACTIVITIES — COUNTERS AND REGISTERS

Investigate each of the circuits listed below, either on a breadboard or by using a simulator, using the suggested ICs. Consult data sheets for connection details.

Build a ripple counter from three or four D-type flip-flops, using 4013 or 74LS112A ICs (negative-edge triggered).

Investigate the action of a 4-stage MSI ripple counter such as the 74LS93 and the 7-stage CMOS 4024 counter.

Run the 4-stage MSI ripple counter at, say 10 kHz, and then use an audio amplifier to 'listen' to the output from each stage.

Design and build a circuit to check the output of a 4-stage ripple counter for glitches. An example is the output '6' which occurs during the transition from '7' to '8'. Connect a set-reset flip-flop to the output of the detector circuit to register when the glitch has occurred.

By using logic gates to detect an output of 1100 (decimal 10), and reset the counter, convert the 74LS93 to a BCD (binary-coded decimal) counter with outputs running from 0000 to 1001.

Design and build a circuit to count in seven stages, from 000 to 110.

Investigate the action of a 4-stage MSI synchronous counter, such as the CMOS 4520 and the 74LS193A (up/down).

Use the circuit you devised in the previous investigation to check for output glitches in a synchronous counter.

Build a shift register from D-type flip-flops.

Demonstate the action of shift registers such as CMOS 4014 (PISO, SISO), CMOS 4015 (SIPO, SISO), 74LS96 (SIPO, SISO, PIPO, PISO), 74LS165A (PISO).

Investigate the action of a SRAM such as the 2114 or 6116. A circuit which can be set up on a breadboard or simulator is shown below. The $\overline{\text{CHIP SELECT}}$ input is held low permanently, and so are all but one of the address inputs. Addressing is controlled by the flip-flop. When this is reset the address is 0 and when it is set the address is 1. The flip-flop is set or reset by briefly shorting one of its inputs to 0 V.

When S1 is not pressed the $\overline{\text{WRITE ENABLE}}$ input is high, putting the chip in read mode. Data already stored at the selected address as a 4-bit word is displayed on the LEDs.

To write data into an address:

- Select the address.
- Set up the data that is to be written by temporarily connecting the data terminals to +5V (1) or 0V (0). On a breadboard, use flying leads.
- Press and release S1. This writes the data into the RAM.
- Remove the connections (flying leads) from the data terminals.
- Read the newly stored data on the LEDs by pressing S1 to put the chip into read mode.

If you select the other address, and then return to the same address, the stored data will still be there. Experiment with storing different values at the two addresses.

Build another flip-flop using the other two NAND gates of IC1. Connect its output to A1. It is then possible to address four words (00, 01, 10 and 11). In these addresses store the data for a traffic-light sequence, assuming D0 is red, D1 is yellow and D2 is green (D3 is not used). Or store the codes for the first four digits of a 7-segment display.

By running through the addresses in order, you can simulate a changing traffic light or make a 7-segment display run from 0 to 3. Devise a circuit to

FIGURE 22.18 A breadboarded circuit for investigating the action of the 2114 SRAM. Similar circuitry may be used for investigating the action of other small SRAMs, such as the 6116.

operate the traffic lights or the 7-segment display automatically.

Try this activity with other small SRAMs with 4- or 8-bit words.

QUESTIONS ON COUNTERS AND REGISTERS

1 Explain what a D-type or J-K flip-flop does and how three such flip-flops may be connected to make a ripple counter.
2 What is the main drawback of a ripple counter? Explain how a synchronous counter overcomes this problem.
3 Describe two logic circuits in which the output of a counter is decoded to trigger a special action.
4 Describe the action of a decade counter with 1-of-10 outputs, and give an example to show how it can be used.
5 Draw a circuit diagram for a serial-in parallel-out shift register, using 4 D-type flip-flops. Explain its action by means of a table.
6 Explain the difference between static and dynamic RAM. What are the advantages and disadvantages of each type?
7 Design and build (or simulate) a 3-bit up-down counter.
8 Explain the principle of a walking ring counter. Name some walking ring ICs and some of their applications.
9 Design and build (or simulate) the decoding circuit for the 6-bit walking ring circuit illustrated in Figure 22.14.
10 Design and build (or simulate) a circuit for operating traffic lights, based on a walking ring.
11 What is flash memory? How does it work? What are its advantages and disadvantages? List three types of electronic equipment that use flash memory.

MULTIPLE CHOICE QUESTIONS

1 A RAM has 11 address inputs and 4 data inputs/outputs. The number of bits that it stores is:
   A 8192.
   B 2048.
   C 44.
   D 4096.
2 To make a decade counter count from 0 to 6 repeatedly, we decode its output to reset the counter at stage:
   A 6.
   B 5.
   C 7.
   D 10.
3 A 4-bit counter has outputs A, B, C and D. To make it flash an LED at one quarter the clock rate, we connect the LED to output:
   A A.
   B B.
   C C.
   D D.
4 A SISO 4-bit shift register has $A = 0$, $B = 1$, $C = 0$ and $D = 0$. How many clock pulses are needed to make its serial output go high?
   A 2.
   B none.
   C 3.
   D 1.

On the Web

Find out more about memory. Select three or four of the newest memory devices and compare them for storage capacity, access time, permanent or temporary storage, maximum number of write operations, power requirements, cost per megabyte, physical size, and any special features.

Display Devices

LIGHT-EMITTING DIODES

Light-emitting diodes, or LEDs, are widely used as indicators and in displays. Although filament lamps and neon lamps are still used, they have been replaced by LEDs in very many applications. LEDs outlast filament lamps, they are cheaper and they require very little current. They are available in a wide range of shapes, sizes and colours.

An LED has the electrical properties of an ordinary rectifying diode. The most important practical difference (apart from the production of light) is that they are able to withstand only a small reverse voltage.

FIGURE 23.1 A typical LED has two terminal wires, anode and cathode. The cathode is often identified by a 'flat' on the rim of the LED body.

For most LEDs the maximum reverse voltage is 5 V, so care must be taken to mount them the right way round.

LEDs are usually operated with a forward current of about 20 mA, with a maximum up to 70 mA, depending on the type. But, if power is limited, for instance in battery-powered equipment, most types of LED give reasonable light output with as little as 5 mA.

The current through an LED must generally be limited by wiring a resistor in series with it (p. 17).

The brightness of an LED is specified in candela (cd) or millicandela (mcd). Typical LEDs are in the range 6-10 mcd. Super-bright LEDs have brightnesses of tens or even hundreds of millicandela, while the brightest types are rated at up to 30 candela. Viewing angle is also important, for some types achieve high brightness by concentrating the light into a narrow beam. This makes them less conspicuous in a display if the viewer is not directly in front of the panel. Standard LEDs have a viewing angle of 120°, but the angle may be as small as 8° in some of the brighter types.

The original LEDs were available only in red, green and yellow, but continuing development has produced a wide range of colours. There are several different tints of red (including ruby red and sunset red) as well as blue and white. Infrared-emitting LEDs are also available.

Bicolour LEDs are useful as indicators. A single bicolour LED can indicate the mode in which an electronic device is operating. For example, on a digital camera a bicolour LED may glow red when it is in recording mode and green when it is in play mode.

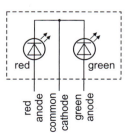

FIGURE 23.2 One type has three terminal wires. The LEDs have a common cathode. A positive voltage applied to either of the other two wires is used to light the corresponding LED.

FIGURE 23.3 In the two-terminal type, the LEDs are connected anode to cathode. Which LED lights depends on which terminal is made positive.

SEVEN-SEGMENT DISPLAYS

When an appliance, such as a clock or a frequency meter, needs to display numerical values, it may use an LED display in the seven-segment format.

FIGURE 23.6 This is how a 7-segment display shows the numerals 0 to 9. The 6 and 9 may sometimes have no tails (segments a and d respectively).

FIGURE 23.4 A seven-segment LED array is used for displaying numerals and letters.

All displays fall into one of two categories, **common anode** or **common cathode**. In the common anode display (below), the anodes of each LED are connected internally to a common terminal. The

FIGURE 23.5 Seven bar-shaped LEDs make up a 7-segment display for numerals 0 to 9. There is also a decimal point, usually on the right, though some displays have a second one on the left.

FIGURE 23.7 With a common-anode display, all anodes are connected to the positive supply and current is drawn through the individual cathodes through series resistors.

FIGURE 23.8 With a common-cathode display, all cathodes are connected to the 0 V rail and current is supplied to the individual anodes through the series resistors.

Seven-segment displays are made in a wide range of types. Most are red but some are green or yellow. The height of the display ranges from 7 mm to 100 mm. Most have a single digit but there are some with 2 or 4 digits. The decimal point is an eighth LED, usually round in shape.

segments are lit by drawing current through their cathodes to current *sinks*. Each segment needs a series resistor.

The common-cathode display (Figure 23.8) has the cathodes of all segments connected internally to a common terminal. Current *sources* are connected to each anode, through a series resistor.

STARBURST DISPLAY

The starburst LED display consists of 14 segments arranged as in the diagram below. It is able to produce any numeral or alphabetic character and many other symbols.

FIGURE 23.9 The coding circuit for a starburst logic display is more complicated than that needed for a 7-segment display.

DOT MATRIX DISPLAY

This often has 35 circular LEDs arranged in 5 rows and 7 columns. The cathodes of all LEDs in the same column are connected to a single line and there is a separate line for each column.

The anodes of all LEDs in the same row are connected to a single line and there is a separate line for each row.

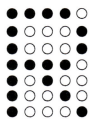

FIGURE 23.10 A dot matrix display is capable of generating a wide range of characters and symbols.

The LEDs forming the character are lit one at a time but in rapid succession. This action is repeated so quickly that each illuminated LED appears to glow steadily and all of them seem to be on at the same time.

Self Test

Draw diagrams to show how the letter of your name can be formed on **(a)** a starburst display, and **(b)** a dot matrix display.

Dot matrix displays require complicated logic circuits to drive them. There are special LSI or VLSI ICs for this purpose, or the task may be performed by a microcontroller. The dot matrix system is ideal for travelling-message displays and other effects in which the characters appear to move sideways or up or down, to flash, or to change size.

LSI and VLSI

Large scale and very large scale integration.

LIQUID CRYSTAL DISPLAY

An LED display is difficult to see if exposed to bright light but has the advantage that it is highly visible in the dark. A liquid crystal display (LCD) has the opposite features. It is difficult to see in dim light but easy to read in direct sunlight.

An LCD consists of liquid crystal sandwiched between two layers of glass. Conducting electrodes are printed on the inside of the front glass, in the format of a 7-segment display. These are almost invisible. Very narrow printed conductive strips connect each segment to its terminal at the edge of the display.

There is a continuous but transparent coating of conductive material on the inside of the back glass, forming the **back plane**.

Normally the liquid crystal is transparent but, if an electric field exists between the back plane and one of the electrodes, the material becomes (in effect) opaque in that region. The connecting strips are too narrow to produce an image. Thus the characters appear black on a light grey ground. The display can be read by reflected light. In darkness, it may have a back light, but the display is not so easy to read under these conditions as an LED display.

The action of the LCD depends on an alternating electric field between the back plane and the electrodes. The segment electrodes must be made alternately positive and negative of the back plane. If this is not done, the electrodes become plated and the device eventually stops working.

The driving circuit requires an oscillator running at about 30 Hz to 200 Hz. This can be built from two

4 Discuss the factors concerned in the transfer of a voltage signal from a radio tuner to a preamplifier.

5 What is meant by a 'crossover network'?

6 Explain why each channel of a stereo system usually has at least two speakers. Describe a typical crossover network.

7 What is a mixer? Describe a simple mixer circuit and explain how it works.

8 Outline the ways in which digital audio signals are stored in a computer and how they may be played out.

OTHER QUESTIONS

9 Describe the structure of a hard disk drive, and how it works.

10 Describe the structure of a colour LCD suitable for use in a mobile phone. (Research on the Web required to answer this.)

11 You are asked to plan (a) a public address system at a railway station, (b) a home theatre, (c) a system to provide background music in a supermarket, (d) an audio/video system for your study/leisure room. Select one of the above and name the electronic units you would include in the system. If possible state manufacturers' names, model numbers, and costs. Describe how to install the system (positioning and mounting the units, cables required, power supply, and so on).

12 Digital technology has made it much easier for people to copy music, images and text. This means that composers, performers and authors may receive little or no payment for the work they have done. What are your views on this? What can be done to make sure that creative people are fairly rewarded?

On the Web

Collect information on stereo and other audio systems and on video systems by visiting manufacturers' sites such as:

www.sony.com,
www.creative.com,
www.lg.com,
www.nikon.com,
www.apple.com, (for iPod)
www.roland.co.jp,
www.yamaha.co.jp,
www.canon.com.

There are also sites that review audio equipment: www.audioreview.com.

Use a browser to find out how these systems work.

MULTIPLE CHOICE QUESTIONS

1 The filter circuits of a stereo system are located at the:
   A power amplifier stage.
   B equaliser stage.
   C preamplifier stage.
   D input stage.

2 A domain on a magnetic disk is:
   A a small magnet.
   B a region of small magnets, all orientated in the same direction.
   C a region in which all the magnetic molecules lie in the same direction.
   D a region in which some of the molecules lie in the same direction.

3 The main advantage of an electret microphone over a crystal microphone is that:
   A its response is linear.
   B it is cheaper.
   C it does not require a power supply.
   D it produces a larger signal.

4 To obtain maximum voltage transfer from a CD player to a preamplifier:
   A the input impedance of the pre-amplifier must be low.
   B the output of the CD player must be high.
   C both the output impedance of the CD player and the input impedance of the preamplifier must be low.
   D the input impedance of the preamplifier must be high.

5 The type of amplifier preferred for the output stage of a stereo system is:
   A Class A.
   B Class B.
   C Class AB.
   D common-emitter.

6 The platters of a hard disk drive may be made from:
   A semiconductor.
   B glass.
   C cobalt alloy.
   D steel.

7 The sensor on the reading head of a disk drive is usually:
   A an electromagnetic coil.
   B a piezo-electric device.
   C a magneto-resistive device.
   D an inductive device.

8 A crossover network:
   A has a push-pull output.
   B is one that has crossover distortion.
   C is a set of filters.
   D gives a flat-topped frequency response.

On the Web

You have been asked to write a 1500-word article on the latest developments in audio-video equipment for a special-interest magazine. The magazine caters for one of these groups: sportsmen and women, music lovers, female teenagers, male teenagers, computer hobbyists, games players, retired persons.

Search the Web for devices likely to be of interest to members of one of these groups. Write an article describing items of interest to them, with technical details if appropriate.

by filtering, or by good design and layout. They are not discussed any further here.

ELECTRICAL CONDUCTION

The types of noise listed next all arise through the nature of electrical conduction. They occur in all circuits and are difficult to reduce:

- **Thermal noise:** This is also known as **Johnson noise**. Electrical conductors all contain charge carriers, most often electrons. Although there is a *mass motion* of carriers when a current is flowing, there is always a *random motion* superimposed on this. Their motion is made random by forces between the carriers and the molecules in the lattice as the carriers pass by. Random motion produces random changes in *voltage*, called thermal noise (Figure 27.2). This is what causes the hissing in the output of an audio circuit. It is similar to the noise made by other random processes such as steam escaping from a steam valve, or water running through shingle on the seashore. The amount of thermal noise depends on the amount of random motion of the charge carriers and this depends on temperature.
- At higher temperatures, the carriers gain extra energy from vibrations of the crystal lattice of the conductor, and so random motion becomes more important.

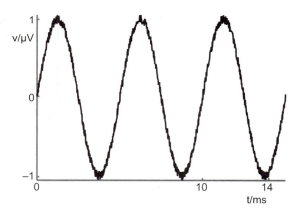

FIGURE 27.2 A 200 Hz sinusoid, amplitude 1 μV, has picked up a 60 nV noise signal, mainly due to thermal noise. The sinusoid now has a randomly spiky form and, if this noisy signal is amplified and sent to a speaker, we hear a hissing sound as well as the 200 Hz tone.

The noise level depends on the absolute temperature, so a conductor needs to be cooled to absolute zero to eliminate this type of noise. Thermal noise

also depends on resistance and on bandwidth. Being completely random, thermal noise has no particular frequency. Or we can take the other viewpoint and say that it contains *all* frequencies at equal amplitudes. Because it contains all frequencies, it is often called **white noise**. If a circuit has a wide bandwidth, a wider range of noise frequencies can exist in it. The total amount of noise in the system is greater. Because thermal noise depends on resistance, it is produced by all components in a circuit, but particularly with those that have high resistance.

- **Shot noise:** With a steady current the *average* number of charge carriers passing a given point in a circuit is constant. But the number passing *at any instant* varies randomly. The random variation of *current* produces another form of white noise called shot noise. Shot noise is a variation of current but the current passes through resistances in the circuit and so becomes converted to variations in voltage. So the effect of shot noise is added to thermal noise.

  Like thermal noise, shot noise includes all frequencies, so the amount of it in a circuit increases with bandwidth. If currents are large (A or mA) the random variations in current are relatively small and shot noise is low. But for currents in the picoamp range, shot noise is *relatively* greater, and may have an amplitude that is 10% or more of the signal amplitude. Shot noise occurs most in semiconductors.
- **Flicker noise:** This is also known as **1/f noise** because its amplitude varies inversely with frequency (Figure 27.3). It is only important when signal frequency is less than about 100 Hz, and may be ignored at frequencies greater than 1 kHz. The cause of it is unknown. It occurs in semiconductors, possibly because of imperfections in the material. It also occurs in valves, where it is possibly an effect of irregularities in the surface of the cathode.

SIGNAL PROCESSING

Processing a signal does not necessarily produce the exact result intended and this can give rise to noise. An example is **quantisation noise**. When an analogue signal is being converted to digital form, a smoothly varying analogue voltage is converted to a stepped digital voltage. The steps introduce an irregularity into the signal which has the same effect as noise. If the conversion has a resolution of 12 bits or more, the stepping usually can be ignored. One bit in 2^{12}, is 1 in 4096, which is an error of only 0.024%.

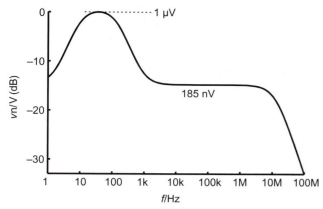

FIGURE 27.3 This graph is an analysis of the common-emitter amplifier on p. 71. It plots the amplitude of the noise signal at the output of the amplifier for frequencies in the range 1 Hz to 100 MHz. Over most of the audio range the noise is 185 nV, due to thermal noise and shot noise. The noise level drops at both very low and very high frequencies because these frequencies are outside the bandwidth of the amplifier, so they are not amplified.

SIGNAL-TO-NOISE RATIO

There are several ways of expressing the amount of noise present in a signal. It is not usually the absolute power of the noise that is important, but rather the power of the noise relative to the power of the signal itself. For this reason, one of the most often used ways of expressing the amount of noise is the **signal-to-noise power ratio**. A ratio of two powers is most conveniently stated on the decibel scale. Because power is proportional to voltage squared, the signal-to-noise ratio is defined as:

$$SNR = 10 \log_{10}(v_s^2/v_n^2) \text{ dB}$$

In this equation, v_s is the rms signal voltage and v_n is the rms noise voltage. The equation has the standard form used for expressing a ratio of powers in decibels.

Example

A signal has an rms value of $v_s = 2.4$ V. The rms noise level is $v_n = 7$ nV. Calculate the signal-to-noise ratio.

$$SNR = 10 \log_{10}(2.4^2/(7 \times 10^{-9})^2) = 171 \text{ dB}$$

Self Test

If the rms noise level is 15 nV and the rms signal level is 350 mV, how much is the SNR?

The SNR is used to calculate the degree to which a circuit, such as an amplifier, introduces additional noise into an already noisy signal. If SNR_{IN} and SNR_{OUT} are the SNRs of the signal at the input and output, the **noise figure** of the amplifier is given by:

$$NF = SNR_{IN}/SNR_{OUT}$$

NOISE REDUCTION

Noise can not be eliminated from a circuit because noise is an inevitable effect of current flowing through components. But it is possible to reduce noise by using the techniques listed below:

- **Avoid high resistances:** Thermal noise is generated when currents pass through resistance and is proportional to the resistance. For low noise, design the circuit so that resistors are as small as possible.
- **Use low-noise components:** Some components are specially manufactured to reduce their noise levels. The BC109 transistor is the low-noise equivalent of the BC107 and BC108 because it has low flicker noise. FETs have no gate current, so no thermal noise is produced. JFETs have less shot noise and flicker noise than BJTs because there is only a small leakage current through their gates. MOSFETs have no shot noise. Many of them have low resistance, so thermal noise is low too. Using FETs in a circuit before the signal is amplified avoids introducing noise that will be amplified by later stages. Often a two-stage amplifier has a MOSFET for the first stage, followed by a BJT in the second stage. Low-noise versions of op amps are another way of avoiding noise.
- **Use low currents:** These result in low thermal noise so it is best to keep current low in the early stages of an amplifier or similar circuit. A collector current of 1 mA is suitably low for a BJT amplifier stage.
- **Restrict bandwidth:** The graph in Figure 27.3 shows that the noise levels in the amplifier fall off at low and high frequencies because they are outside the bandwidth of the amplifier. Restricting the bandwidth can further reduce noise. In this particular amplifier the upper -3 dB point is at 11 MHz. For many applications there is no need to pass signals of such high frequency. In these cases, reducing the high-frequency response will eliminate much of the noise, without affecting its performance in other ways.

The bandwidth of radio frequency amplifiers needs to cover only the carrier wave frequency and the sidebands which occur about 15 kHz to either side of this. Restricting the bandwidth to such a narrow range means that noise is very much reduced. In general, noise may be reduced by restricting the bandwidth of a circuit to just that range of frequencies it needs to do its job.

in their quietest passages. Also the higher harmonics of all instruments are of low power. In a compander, all of these are boosted relative to other more powerful signals. The loudest instruments remain the loudest and the quietest remain the quietest but the range

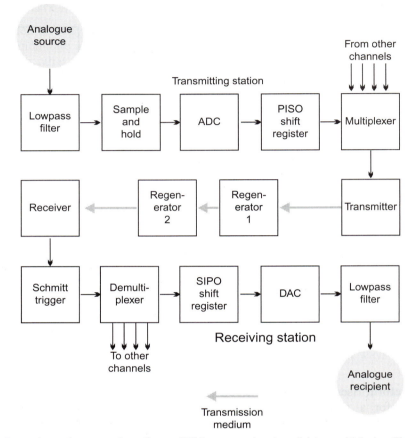

FIGURE 28.10   Transmitting analogue data over a long-distance PCM system, using time division multiplexing. The transmission medium could be cable, optical fibre or radio. A system for transmitting digital data would be similar, but would replace the ADC and DAC by a coder and a decoder at each end of the system (for 2-way communication). The coder and decoder may be combined into one device, known as a codec. Instead of a hardware codec, the same functions may be taken over by a codec program running on a computer.

- Digital signals are highly immune from noise. The signal shown in Figure 28.12 is noisy but there is no problem in recognising it as a pulsed signal.
- Distorted or noisy pulses can easily be squared up with a Schmitt trigger. This is not true of analogue signals.

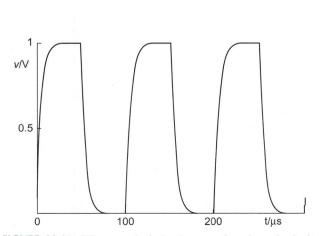

FIGURE 28.11   When a pulsed signal passes through a circuit, its shape is distorted by stray capacitance.

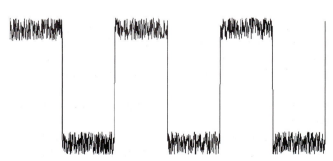

FIGURE 28.12   A high level of noise does not prevent the pulses from being readable.

- On long routes, regenerators can restore the shape of the pulses. This is not so easy with analogue signals, which need more repeater stations along the route.
- Digital signals are suited to processing by computers and other digital logic circuits. This makes coding, decoding, error checking, and multiplexing much easier.

NETWORKS

Many different types of system are in use for communication between stations. The main aims of these systems are to:

- Transfer information as soon as possible and as quickly as possible.
- Transfer information without error or loss.
- Make maximum use of the transmission medium, both in terms of *time* (minimising idling time), and *bandwidth* (transmitting many channels simultaneously).

The final point is important because submarine cables, communications satellites and other hardware (see next three topics) are expensive to build, maintain, and operate.

We will look at ways of connecting information sources with recipients, starting with the simplest systems and going on to world-wide networks. It is essential that communicating stations must 'speak the same language', in the sense of using compatible coding and decoding procedures, compatible multiplexing and demultiplexing procedures and, in general, make certain that messages transmitted from one station can be received and interpreted by the recipient. To this end, a wide range of **standards** and **protocols** have been devised by international bodies. We will look at a few typical and widely used examples of these.

RS-232

This is a standard for cable communication between two computers. The term 'computer' in this account includes not only personal computers but any computer-like equipment such as data-recorders and modems.

Data can be output from a computer either in **parallel** form or in **serial** form. A parallel connection has a line for each bit in the data words. For example, in an 8-bit system there are 8 parallel data lines, plus ground (0 V) lines and a few other lines for handshaking between the stations. This kind of link is

used for sending signals from a PC to an attached printer. The Centronics interface is an example. The advantage of parallel connection is that a whole byte of information can be transmitted in one operation. The disadvantage is that it is limited to cables a few metres long, Usually, transmitter and receiver are in the same room. The RS232 ('RS' stands for 'Recommended Standard') connection is an example of a serial connection.

The complete RS232 connection uses a 25-way connector, and most of the lines are in use. However, this provides features such as a secondary channel and timing signals that are not generally used. Timing is necessary for **synchronous** data transmission, in which the transmitter and receiver must be kept in step. Most users of RS232 employ **asynchronous** data transmission, as described below.

The simpler RS232 connection, which is the one most commonly used, has a 9-pin connector. The directions in which signals are transmitted are illustrated in the diagram. This shows a typical case, in which a computer is communicating with a modem.

FIGURE 28.13 A computer can be connected through an RS232 cable to a modem, and then to the telephone network.

Three of the nine lines are concerned with carrying the data:

TXD (or TD): transmits serial data from computer to modem.

RXD (or RD): receives serial data from modem to computer.

SG: signal ground, the common return path for TXD and RXD signals.

Four of the lines are concerned with handshaking. These are the lines by which the computer and modem coordinate their actions:

RTS: request to send. The level on this line goes high when the computer is ready to transmit data to the modem along the TXD line. This is the first stage of handshaking.

disassembler or **PAD**. The packet begins with a block of data or **header** which contains such information as:

Synchronising bits.
Address of origin.
Destination address.
Sequence number.
Error checking code.

The sequence number is used when the message extends over two or more packages. It might happen that the different sections of a message are sent along different routes, so they arrive at their destination out of order. The PAD at the destination uses the sequence numbers to combine the sections of the message into a single message, with all its parts in the right order.

Following the header comes the main message and then the **footer** that concludes the transmission. Header and footer are stripped off at the receiving end and only the message reaches the recipient.

Networks that employ packet switching may have several different configurations. A **bus network** has the terminals spaced out along its length, each communicating with the terminals on either side of it. There are also ring networks and star networks. The transfer of data in the network is performed according to agreed rules, known as a **protocol**.

TCP/IP

A widely used protocol, used on the Internet and many others, is known as **Transport Control Protocol/Internet Protocol**. The Internet is a worldwide network. It has grown enormously since it was first developed in 1969 for military purposes. It has incorporated several other networks and certain networks, such as Compuserve, operate in conjunction with it. The vast number of users attracted by the World Wide Web, which runs on the Internet, has contributed to its expansion.

At present, the Internet uses a protocol known as **IPv4**, which is Version 4 of the Internet Protocol. Its packet-switching format is described in the previous section. One limitation is that the address field is only 32 bits long. This gives 2^{32} possible addresses, which is over 4 thousand million, but this is insufficient to meet demand. In writing the address of a computer on the Internet, the 32 bits are divided into 4 bytes of 8 bits each. These are then written out as 3-digit decimal numbers, each in the range 000 to 255, separated by full-stops, generally known as **dots**.

For example, an address might look like this hypothetical example:

123.111.222.001

In the future, if IP Version 6 comes into operation, addresses will have 128 bits.

Transport Control Protocol is concerned with ensuring that a packet arrives at its destination free from error. If a package is lost or arrives in corrupted form, TCP returns a message to the sender asking for a repeat transmission. If a message in broken into several packets, which arrive out of order, TCP places them in the correct order.

TCP also controls the rate of data transmission. If too many packets are being lost, perhaps because of bad reception or overloading, the rate of transmission on the network is reduced to allow the system to cope better.

An address that is a string of 12 digits is not easy to remember. Keying in such an address is prone to error.

To overcome this problem and to help make the system more user-friendly, the Internet has introduced a system of **domain** addresses. Each computer on the Internet has, as well as its numerical address, an address in the following format:

newnespress.com

The lower level domain name is 'newnespress', which is owned by the publishers of this book. The upper level domain name is 'com'. This is three-lettered and indicates the nature of Newnes Press, which is a commercial body. Other upper level domain names are 'org' for a non-commercial organisation, 'ac' or 'edu' for an academic organisation such as a university, 'gov' for a governmental organisation, and 'net' for a network. Commercial sites located in certain countries (such as the United Kingdom) use the name 'co' instead of 'com'.

The domain name may also have a two-lettered third part, which indicates the country in which the computer is sited. For computers in the USA, this is generally omitted, but examples for other countries are 'uk' for United Kingdom, 'au' for Australia and 'jp' for Japan.

Sites that are connected on the World Wide Web have their domain names prefaced by 'www'.

As might be expected, there are organisations in each country supervising the issuing of domain names. A table of domain names is updated continually and held on the routing computer (or **routers**) in the network. Given any domain name, the router can use its **look-up table** to find the corresponding numerical address and thus begin the process of forwarding the packet to its destination.

At a router, packets are continually being received and checked using complex error-checking procedures. The packet is then sent on to the next router on its way to its destination. If the shortest route is heavily loaded, the router is able to select an alternative route for the packet

QUESTIONS ON TELECOMMUNICATIONS

1 Match the stages of the diagram in Figure 28.1 against the stages of transmission of information by: a smoke signal, semaphore, a video surveillance system in a supermarket, and any other named system of your choice.
2 Name some sources of information that make up the first stage of a telecommunications system. In what form(s) does the information exist?
3 Explain, giving examples, the functions of transducers in information systems.
4 What is modulation? Explain the difference between amplitude modulation and frequency modulation.
5 Compare the advantages and disadvantages of amplitude modulation and frequency modulation.
6 What is a sideband? In what way does the production of sidebands affect the bandwidth of a modulated carrier?
7 Describe how an analogue signal is transmitted and received using pulse width modulation.
8 Compare pulse width modulation with pulse position modulation.
9 Describe how a binary message is sent using frequency shift keying.
10 Explain what is meant by the amplitude, rise time and decay time of a signal pulse.
11 Describe the transfer of information over an RS232 cable with 9 wires.
12 The Association for Science Education gives its Internet address as:

   www.ase.org.uk

   Explain the meaning of the four parts of this address.
13 What are the meaning of these terms: protocol, router, packet switching.
14 Describe the structure and operation of a local telephone network.
15 What is frequency division multiplexing? What are its advantages?

16 An SSB voice signal requires a bandwidth of 3.4 kHz. If it is modulated on to an 80 kHz carrier, what frequencies does it contain?
17 What is the bit error ratio of a transmission? How is it measured?
18 What is jitter? Name one of its causes. How can it be reduced?
19 Draw diagrams to show the waveforms of these pulse signals: (a) return to zero, (b) non-return to zero, (c) non-return to zero mark.
20 What is the advantage of an RZ system over NRZ systems? What is its disadvantage?
21 Draw a diagram of the wave form of this signal, encoded as NRZ mark: 001110110.

MULTIPLE CHOICE QUESTIONS

1 An AM signal has a modulation depth of 20%. This means that:
   A the signal frequency may vary by as much as 20%.
   B the modulated carrier amplitude varies by as much as 20%.
   C the signal frequency is 20% of the carrier frequency.
   D the amplitude of the signal is 20% of the amplitude of the carrier.
2 A modulation system in which there are pulses of constant amplitude and length but with varying timing is known as:
   A PPM.
   B PWM.
   C PCM.
   D PDM.
3 One of the systems used for transmitting binary data is:
   A PCM.
   B PWM.
   C PDM.
   D PAM.
4 Of the systems listed below, the one most subject to interference from spikes and noise is:
   A FSK.
   B AM.
   C FM
   D PPM.
5 When a pulse signal oscillates before settling to its new level, the effect is called:
   A overshoot.
   B the time constant.
   C ringing.
   D oscillations.

6 A carrier with frequency 250 kHz is amplitude modulated with a signal of which the maximum frequency is 8 kHz. The bandwidth of the signal is:

   A 258 kHz.

   B 16 kHz.

   C 8 kHz.

   D 242 kHz.

7 A system operates with even parity. The seven bits of a byte are 0110101. When the parity bit is included the byte is:

   A 10110101.

   B 01101010.

   C 00110101.

   D 01101011.

8 What type of signal is shown by the drawing shown on the right?

   A NRZ.

   B RZ.

   C PWM.

   D FSK.

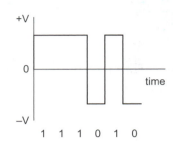

9 If a system has jitter:

   A the timing of the pulses is irregular.

   B the rise time of the pulses is too slow.

   C it has a high bit error ratio.

   D there is a fault in its codec.

Cable Transmission

Telecommunication by cable is an extension of the passing of signals directly from the circuit to external conductors. In a wired intercom, for example, the sound signal is carried along a pair of wires from the master station to the remote station. When the master station is trasnmitting, the remote station is in essence just a speaker at the end of a long lead. The signal is unmodulated. The distance between transmitter and receiver is limited to a few hundred metres. Communication between a telephone in the home and the local exchange falls into the same category.

Most cable communications differ from the above in one or both of the following ways:

- The distance between transmitter and receiver is greater.
- The signal is modulated on to frequencies much higher than audio frequency.

These factors have important effects on transmission. To understand these we first look at the types of cable used and also examine some of the properties of transmission lines.

TYPES OF CABLE

A wide range of cables is available, but the two types most commonly used are the twisted pair and coaxial cable.

The **twisted pair** consists of two insulated wires twisted around each other. Because the wires are unshielded (that is, they have no metallic sheath around them), there is capacitance between the two wires and between each of the wires and ground or any other nearby conductors. Because they are twisted around each other, the capacitances of each wire to ground or to external conductors is balanced. Any capacitive effects are common mode and have little effect on the signal. The cable is said to be **balanced**.

(a)

(b)

FIGURE 29.1 The two types of cable most often used are (a) the twisted pair, and (b) coaxial cable.

Common Mode
Affecting both wires equally.

The induction of signals produced by external magnetic fields is minimised by twisting. In any circuit which contains one or more loops, an external magnetic field passing through each loop induces an e.m.f. in that loop. But in a twisted pair, the direction of induced e.m.f.s in the wires is reversed in adjacent loops (see below) so they cancel out.

⊙ magnetic field

induced emf

FIGURE 29.2 The loops of the twisted pair are threaded by a magnetic field increasing in a direction upward through the paper of the page. By Fleming's Left-hand Rule, this induces an e.m.f. in a clockwise direction in each loop. Because of the twisting, these e.m.f.s cancel out in adjacent segments of the wire. Changes in direction and intensity of the field affect all loops equally.

Coaxial cable is widely used for carrying radio-frequency signals. The signal is carried in the inner conductor, which may be a single copper wire or may consist of a number of narrower copper strands. Stranded wire is used when flexibility is important. The inner conductor is insulated from the outer conductor by the dielectric, a layer of polythene or polyethylene.

The outer conductor is made of copper and is usually braided to give it flexibility. This is connected to ground so it serves to shield the inner conductor from external magnetic fields. It also prevents magnetic fields produced in the inner conductor from spreading to other nearby circuits. Because there is no capacitance between the inner conductor and external conductors and, because the inner conductor is shielded from external magnetic fields, a coaxial cable is described as **unbalanced**.

TRANSMISSION LINES

When a twisted pair or coaxial cable is being used for transmitting signals over a large distance, or the signal has high frequency, or both circumstances apply, we have to consider the properties of the medium in more detail. We regard this not as a mere length of conductor but as a **transmission line**. There is no sharp distinction between a simple lead, such as a speaker lead, and a transmission line, such as a submarine cable. It is just that, as length increases, or as transmission frequency increases, the distinctive properties of the line become more and more important.

Not all transmission lines extend for kilometres under the oceans. In microwave transmitters and receivers the frequency is so high (up to tens of gigahertz) that the wavelength of the signal in the conductor is only a few tens of millimetres. Even conductors joining one part of the circuit to another part need to be treated as transmission lines. At the very highest frequencies the mechanism of propagation of the signal becomes rather different. We use **waveguides** (box-like metal channelling or metal strips) instead of ordinary cable.

A tranmission line may be considered to be an electronic circuit in its own right. The diagram in Figure 29.3 represents a balanced transmission line by its equivalent electronic components. The resistors R represent the resistance of the copper conductors. In reality, the resistance is spread along the whole length of the wires, but here we represent it by individual resistors. The resistors are the **lumped equivalents** of the evenly distributed resistance of the wire.

FIGURE 29.3 The lumped equivalent of a transmission line. In theory, the line consists of an infinite number of units like these three, spaced equally along the line.

If the cable is a short one carrying audio signals, resistance is the only property of the line that is important. Ohm's Law is all that is needed for analysing the circuit. With short runs of a suitable gauge of copper wire, even resistance may be ignored.

Inductance is another factor that adds to the impedance of the cable. Inductance is usually associated with coils, but even a straight wire has inductance (consider the wire to be part of a one-turn coil of very large radius). In the figure, this too is represented by its lumped equivalent, the two inductors.

Capacitance between the wires is represented by the capacitors C, joining the wires at regular intervals. The size of the capacitance depends on the spacing between the conductors and the material (dielectric) between them. Usually the dielectric is plastic, often PVC in the case of a twisted pair, or polythene in a coaxial cable.

Finally, there is always a small leakage current through the insulation or dielectric, represented by the conductances, G.

Using lumped equivalents reduces the transmission line to a relatively simple 'circuit', which can be analysed by the usual techniques. For instance, we know that the reactance of capacitors becomes very low at high frequencies and it is easy to see from the diagram that capacitance becomes more important than conductance. Conversely, the reactance of inductors becomes very high at high frequencies, so inductance becomes more important than resistance. As a result, conductance and resistance may be ignored at high frequencies (see Figure 29.4).

FIGURE 29.4 At high frequencies, resistance and conductance may be ignored, leaving only inductance and capacitance.

At high frequencies, the lumped equivalent circuit consists of only capacitance and inductance, as illustrated above. This arrangement of capacitance and inductance has the properties of a lowpass filter. Signals passing along the line lose relatively more of their high-frequency components. This has the effect of making pulses more rounded. They lose their sharp edges. They become more spread out. If this occurs, adjacent pulses may merge, causing errors in the signal at the far end of the line.

The dominance of capacitance and inductance at high frequencies leads to another effect. Inductance and capacitance store and release energy but they do not absorb it (p. 57). *In theory*, there is no loss of energy or power along a transmission line with a high-frequency signal. *In practice*, some of the energy escapes as electromagnetic radiation (in twisted pairs, less in coaxial cable), or as heat generated in the conductors when current passes through them.

Summing up, there is a fall in signal power along the line. This can be ignored in short lines, which we then consider to be **lossless**.

CHARACTERISTIC IMPEDANCE

The most important characteristic of a transmission line is its characteristic impedance, Z_0. This is defined as the ratio between the voltage and current (v/i) at the input of a line of infinite length.

If the impedance of the signal source is equal to Z_0, all of the power generated by the source is transferred to the line. Conversely, if there is an impedance Z_0 at the far end of the line, all the power of the signal is transferred to this. It follows that a lossless line transfers all the power from a source to the receiver. But, for this to happen, the source and receiver impedances must be **matched** (equal) to the line impedance.

If the impedance at the far end of the line is not equal to Z_0 (including when it is an open circuit, or when there is a short circuit) part or all of the energy of the signal is reflected back along the line. This causes standing waves to form in the conductors, and less power is transferred to the receiver. For efficient transmission of signals, this is to be avoided by careful matching of impedances.

The characteristic impedance of a cable at high frequencies depends on the inductance and capacitance per unit length of cable. It is independent of frequency, provided that frequency is high. In other words, the cable behaves like a pure resistance, which remains fixed, whatever the frequency.

In practice, cables are made with standard characteristic impedances. Transmission hardware has input or output impedances that match these standards. The most commonly available value of Z_0 for coaxial cable is 50 Ω, another common value being 75 Ω. Twisted pairs have standard impedances of 300 Ω or 600 Ω.

With lower frequencies, the impedance varies with frequency. The range of frequencies that the line can transmit without significant loss of power is called its **bandwidth**. The bandwidth extends from the lowest frequency to the highest frequency at which the output signal is less than 3 dB below the input signal. The graph below shows a simulation of a transmission line consisting of a twisted pair of copper wires with characteristic impedance of 600 Ω and length 200 m.

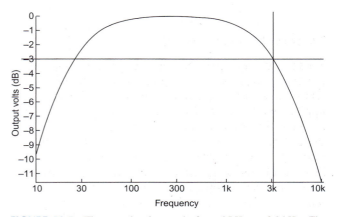

FIGURE 29.5 The pass band extends from 25 Hz to 3.0 kHz, The bandwidth is 3000 − 25 = 2975 Hz.

The amplitude of the signal across the load at the receiving end is plotted against frequency. Both are plotted on logarithmic scales, and the amplitude is scaled in decibels. The line at −3 dB indicates the half-power level.

Signal strength rises above half power at 25 Hz and falls below it at 3 kHz. The bandwidth of the system is defined as the difference between these two frequencies, which is 2972 Hz.

LINE DRIVING

For analogue audio signals, a pair of centre-tapped transformers may couple the transmitter and receiver to the line, as illustrated in Figure 29.6. The taps are grounded so that when one line of the twisted pair is positive of ground, the other is negative of ground by exactly the same amount. The line is **balanced**.

FIGURE 29.6   Source and load may be transformer-coupled to a balanced transmission line to minimise EMI.

This minimises electromagnetic interference because such interference will affect both lines equally, making one more positive and the other less negative. The interference is common-mode and has no effect at the receiver.

If the source circuit has a high output impedance, a buffer is needed between it and the transformer to match the source to the line.

Audio signals may be fed directly to the line using a **line driver**. A wide range of these is available as integrated circuits, which usually have several drivers and receivers in the same package. Each driver is a differential amplifier with low output impedance to match the impedance of the line and with high slew rate and high bandwidth. It has two outputs (see below) which are in antiphase for connection to the two lines of a balanced transmission line.

Self Test

What is the meaning of *slew rate*?

The receiver is a differential amplifier and the two wires of the transmission line are connected to its (+) and (−) inputs.

FIGURE 29.7   The essentials of using a line driver and receiver to send data along a twisted pair, or any other form of balanced line. The line is terminated in a load resistor equal to the characteristic impedance of the line.

The receiver has high slew rate, high bandwidth and high common-mode rejection. Although most types of driver and receiver are intended for use with balanced lines, they are also able to operate on unbalanced lines, in which one path is grounded and the signal passes along the other.

For digital and pulsed signals, the line driver is usually a logic gate, often NOT or NAND. TTL buffer gates can be used, as in the diagram below, though special line driver gates are preferred, because these have higher slew rates and higher bandwidth. CMOS drivers are used in low-power circuits.

The receiver is also a logic gate, sometimes a NOT gate. It may have Schmitt trigger inputs to 'square up' the distorted incoming pulse and to reduce the effects of noise on the line.

FIGURE 29.8   A basic circuit for using a line driver and receiver to send data along an unbalanced line. The line is terminated in a load resistor equal to the characteristic impedance of the line, usually 50 Ω or 75 Ω in the case of a coaxial cable.

A wide range of ICs is available for interfacing logic circuits to RS232 cables. The MC1488P, for example, contains four RS232 line drivers. Three of these function as NAND gates and the fourth as a NOT gate. The IC requires a power supply of up to +15 V and −15 V. It converts input from a TTL circuit into the equivalent RS232 levels. As these are inverting gates, the conversion includes inversion. Thus, an input of 0 V (TTL logic low) produces an output of +15 V, while an input of 5 V (TTL logic high) produces an output of −15 V.

The MC1489P has four NOT gates for use as line receivers. They have Schmitt trigger inputs to square up the incoming pulses. The IC operates on 5 V to 10 V and can accept inputs of up to ±30 V. Its outputs are the corresponding standard TTL levels of 0 V and 5 V.

Because most equipment connected by RS232 operates on TTL voltage levels, the more recent line driving and receiving ICs operate on +5 V and +3 V (many recent computers operate on this lower voltage). There is no need for the ±12 V (or higher) supply that is usually associated with RS232 systems. Instead, the IC contains a circuit unit that generates the higher voltages from the low supply voltage. This greatly simplifies construction of RS232 interfaces.

CROSSTALK

When several lines are close together, the current passing along one line may cause current to flow in one or more adjacent lines. This may be the result of capacitance between the two lines or the effect of inductance. The effect is known as **crosstalk**, and often occurs in multicored cables and in the ribbon cables used for connecting computers to peripherals. It may be eliminated in multicored cables by having each of the cores enclosed in a separate earthed shield, though this makes the cable considerably more expensive.

Connecting alternate lines to ground helps to reduce crosstalk in ribbon cables. Crosstalk also occurs between adjacent tracks on circuit boards, particularly if the tracks run side-by-side for appreciable distances. It is avoided by running grounded tracks between the active ones. The size of the current induced by a magnetic field is proportional to the *rate of change* of the magnetic flux. This in turn is proportional to the rate of change of the current generating the field. Therefore, induced crosstalk is more serious at higher frequencies.

The effect is the same for capacitative crosstalk because capacitance decreases with frequency and high-frequency signals pass more readily across the capacitative barrier.

In data signalling at high rates, the pulses have very rapid rise-times, and these result in the most serious crosstalk problems. Spikes several volts high may be generated on an adjacent line at the beginning and end of each pulse. The effect can completely disrupt the signal on the adjacent line. With many types of line driver the rise time can be adjusted so that it is no more rapid than necessary, so minimising crosstalk.

ACTIVITY — TRANSMISSION LINES

Set up a circuit in which a sinusoidal voltage source feeds a signal through a source resistor into a transmission line, and there is a load resistor at the far end.

The transmission line could be a 100 m length of twisted pair or coaxial cable. The resistor values are equal to the characteristic impedance of the line, which is usually given in the retailer's catalogue. This circuit may be breadboarded, or set up with simulation software.

Use an oscilloscope or the corresponding software feature to examine voltage signals at the input and output of the line. Measure the amplitudes of the signals and calculate their maximum power, using $P = V_{rms}^2/R$.

Run the test with various load resistors in the range 10 Ω to 1 kΩ to find which resistor obtains the greatest amount of power from the line. Try this for sinusoidal signals for a range of frequencies from 1 MHz up to 20 MHz.

Here are some hints for setting up the tests. At high frequency, $Z_0 = \sqrt{(LC)}$. In this equation, L and C are the inductance and capacitance per metre. Typical values for a twisted pair are $L = 600$ nH and $C = 50$ pF. Applying the equation, these values make $Z_0 = 109.5$ Ω, independent of the length of the line.

If the simulator requires a value for R, a typical value is 0.17 Ω per metre. Obtain values for other types of line from data sheets.

To calculate the delay time taken for a signal to pass along the line, begin with the velocity of the signal in free space, which is 300×10^6 m/s. The velocity is less than this in a conductor, usually between 0.6 and 0.85 of the free space velocity. A typical delay for a 100 m line is 475 ns.

Replace the sinewave generator with a square-wave generator and investigate the effect of the line on the shape of the pulses at various frequencies, with matched and unmatched loads.

ACTIVITY — TRANSFORMER COUPLING

Set up the circuit opposite using 1:1 line-isolating transformers. Apply an audio-frequency sinusoid at one end (through a source resistor) and use an oscilloscope to examine the signal across the load resistor.

Calculate the power of the input and output signals for various combinations of source and load resistors in the range 50 Ω to 10 kΩ.

This circuit can also be run as a simulation.

ACTIVITY — LINE DRIVERS AND RECEIVERS

Use data and circuits from the manufacturers' data sheets to set up a 100 m balanced transmission line. For audio signals (sinusoids in the audio frequency range) use drivers such as the SSM2142 and receivers such as the SSM2143. Use an oscilloscope to investigate the transmisson of signals along the line.

Set up a 10 m unbalanced coaxial transmission line and use a pair of RS232 driver and receiver ICs to transmit digital data along the line at up to 20 kHz.

The angles at which these effects occur depend on the wavelength (colour) of the light. This causes a beam of ordinary white light to spread into its component colours. The diagrams assume that the light is monochromatic so the ray remains narrow.

In Figure 30.3 below, light is passing along inside a narrow cylinder of glass. When it strikes the glass-air surface it is not able to emerge into the air. This is because θ_1 is greater than c. The ray is totally reflected. It passes along inside the cylinder and may be totally reflected many more times before it emerges from the other end. This is the basis of the **optical fibre**. Even if the fibre is curved, as in the figure, the light travels from one end to the other without emerging.

FIGURE 30.3 A ray of light travelling in a glass fibre is totally reflected when it strikes the surface of the glass. If the fibre has slight bends in it, the ray is conducted along the fibre without loss.

It may sometimes happen that a fibre is too sharply curved. The θ_1 is less than c and the light emerges. Light also emerges if there are scratches in the surface of the glass, producing local regions where θ_1 is less than c.

A practical fibre-optic cable consists of a narrow fibre of glass, the **core**, surrounded by a layer of glass of lower optical density, the **cladding**.

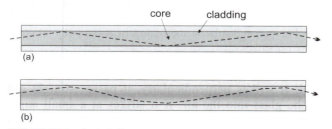

core   cladding

(a)

(b)

FIGURE 30.4 Optical fibre usually has (a) a core of high refractive index surrounded by a cladding of low refractive index; or (b) the refractive index decreases gradually from the centre out. In (a) the ray is totally reflected every time it strikes the boundary. In (b) the ray is gradually redirected back to the centre of the core whenever it deviates into the outer region.

Note that the reflecting surface is the inner surface of the cladding, so it is protected from becoming scratched.

In **graded index** fibre, the refractive index decreases gradually from the centre of the core outward. The ray is bent gradually as it enters the outer regions and is gradually returned to the central region. Its path along the fibre is smoothly curved

instead of being a sharp zigzag. This type of cable produces less distortion of light pulses and is better for long distance communication.

The glass used for fibre optic cable is of very high purity and clarity, prepared in a clean environment. The loss of signal due to absorption in the glass is about 0.15 dB (3.5%) per kilometre. Cheaper cables made from polymer are used for short-distance runs (up to 100 m). They have a higher absorption rate. Fibres of either type are enclosed in a tough plastic sheath for protection.

LIGHT SOURCES

The light signal is provided by a light-emitting diode or a solid-state laser, both of which produce monochromatic light. The source is enclosed in a metal housing which clamps it firmly against the highly polished end of the fibre. LEDs are inexpensive and reliable but their light output is low. The light output is proportional to the current. The intensity (amplitude) of the light signal is easily modulated and this is the most commonly used modulation system for LEDs.

Solid-state lasers are more expensive than LEDs and require more complicated circuits to drive them, but their power is greater. Their output may be modulated in intensity, frequency or phase, and also by the angle of polarisation. Like LEDs, they are mounted in a metal housing.

At present most of the LEDs and lasers operate in the infrared region but, with further research, operation is being extended into the visible spectrum.

Because of the very high frequency of light waves, it is possible to modulate the light signal at very high rates, measured in terahertz (1 THz = 10^{12} Hz). Consequently, the bandwidth of optical channels is around 100 times wider than that of radio channels. Such a wide frequency range provides enormous bandwidth for stacking simultaneous transmissions on a single channel. There is room to carry 100 000 voice channels over a single optical fibre channel. This is one reason why optical fibre is becoming more and more widely used for telecommunications.

LIGHT RECEIVERS

The devices used for detecting the signal and converting it to an electrical signal need to have very high sensitivity and the ability to operate at high frequencies. The most commonly used receivers are avalanche photodiodes, PIN photodiodes and phototransistors.

Avalanche photodiodes (APDs) are operated with reverse bias. When stimulated by light, a few electron-hole pairs are generated. These move in the electric field, striking other atoms in the lattice and generating more electron-hole pairs. The effect is similar to that in an avalanche Zener diode. The result is a rapid increase in the number of electron-hole pairs and a rapid increase in the current passing through the diode. The response time is about 200 ps, so the device is suited to high-frequency operation.

> **Intrinsic Material**
> Pure, undoped semiconductor, with low conductivity.

PIN photodiodes are so-called because they have a layer of intrinsic (i-type) material between the n-type and p-type material. This widens the gap between the n-type and p-type so that the capacitance between them is reduced. With reduced capacitance, the PIN diodes are more suited to operate at high frequencies. The wide gap also increases the absorption of light at the junction, so increasing sensitivity.

> **Self Test**
> What charge carriers are present in the depletion region of a reverse-biased diode?

PIN diodes are reverse-biased in operation, and light causes the generation of electron-hole pairs. In a circuit such as that below, the increased current passes through the resistor and produces an increased voltage across it. This is amplified to produce the electrical signal.

FIGURE 30.5 A basic circuit for converting an AM optical signal into an AM electrical signal.

A **phototransistor** is enclosed in a transparent package or in a metal can with a transparent lens-like cap. It is connected into a circuit in the same way as an ordinary npn transistor, except that it is not necessary to make any connection to the base. When light falls on the transistor it generates electron-hole pairs in the base layer. A current then flows from base to emitter, which is equivalent to the base current of an ordinary BJT.

Because of the transistor action, a much larger current flows from collector to emitter. Their amplifying action makes phototransistors more sensitive than photodiodes, but they are less able to operate at high frequencies.

ADVANTAGES OF OPTICAL FIBRE

The main advantages of optical fibre are:

- **Bandwidth:** optical systems have a higher bandwidth than any other system of telecommunication. Many more signals can be frequency multiplexed or time multiplexed on to a single channel. This allows the development and implementation of highly complex stacking systems.
- **Cable cost:** the cost of optical fibre is greater than that of copper cable, but it is falling. Offset this against the fact that, over long distances, repeater stations can be up to 160 km apart. Often they are much closer than this because the power for the repeater unit is supplied along a pair of copper wires running in the cable alongside the fibres. This limits the distance between repeaters to 10 km. Local power supplies are used if the distance between repeaters is greater than this. Even so, optical fibres give longer runs compared with copper cables, which need repeaters every few kilometres.
- **Immunity from EMI:** unlike copper cable, optical fibre is immune from the effects of external electromagnetic fields. It is particularly suited to industrial applications where electrical machinery and heavy switching generate strong interference. There is no crosstalk between fibres bundled in the same cable.
- **Safety:** faults in copper cabling may generate excess heat, which may lead to fire. This is much less likely to occur with optical fibre.
- **Security:** surveillance equipment can be used to read, messages being transmitted on copper cables. This is not possible with optical fibre.
- **Corrosion:** unlike copper cable, optical fibre does not corrode.

QUESTIONS ON OPTICAL TRANSMISSION

1 Describe what happens when a ray of light travelling in air meets an air-to-glass surface.
2 Describe what is meant by critical angle.

3 What is graduated index optical fibre and how does it work?

4 Describe how an LED is used as the light source for optical fibre.

5 Describe the action of an avalanche photodiode as an optical fibre receiver.

6 What are the advantages of optical fibre as a medium of telecommunication?

MULTIPLE CHOICE QUESTIONS

1 Total reflection occurs when a ray of light:
A passes from air to glass
B is refracted away from the normal
C strikes the surface at an angle less than the critical angle
D strikes the surface at an angle greater than the critical angle.

2 It is important not to bend an optical fibre too sharply because:

A it may break
B it alters the refractive index
C light may escape at the bend
D the sheathing plastic may split.

3 The type of receiver most suited for high-frequency reception from an optical fibre is:
A a PIN photodiode
B an avalanche photodiode
C a laser diode
D a phototransistor.

On the Web

Find out more about optical fibre, light sources and receivers. Details to look for include transmission frequencies, and bandwidths. Try keywords and phrases such as optical fibre, infrared, and infrared laser.

Radio Transmission

THE ELECTROMAGNETIC SPECTRUM

When an electron is accelerated, it radiates energy in the form of **electromagnetic waves**. We can think of the waves spreading through space like ripples on a pond, but in three dimensions. The wave fronts are spherical and continuously expanding.

Electromagnetic waves include (in order of increasing wavelength): gamma rays, X-rays, ultraviolet, visible light, infrared, microwaves, and radio waves. All are the same type of radiation and have similar properties. For example, they all travel through free space with the same velocity, 300×10^6 m/s. Their wavelength (λ) is related to frequency (f) by the equation:

$$\lambda = \frac{3 \times 10^8}{f}$$

They can be reflected, refracted, and diffracted, for example, and they can be absorbed by the medium through which they are travelling.

There are differences. For example, gamma rays and X-rays can penetrate dense matter such as metals, but radiation of longer wavelength can not. These radiations and also ultraviolet are ionising radiations. They cause electrons to be knocked free of the atoms of materials as the radiation passes through. They can also be very damaging, possibly lethal, to living tissues. They produce the ionised layers of the upper atmosphere that are so important in radio transmission.

Radio waves have the longest wavelengths of all types of electromagnetic radiation. The shortest wavelengths (100 mm to 1 mm) are found in the Extremely High Frequency band (30 GHz to 300 GHz). The longest wavelengths (1000 km to 10 000 km) are in the Extremely Low frequency band (30 Hz to 300 Hz). All are used for telecommunications of various kinds.

The two other wavebands used in telecommunications are visible light and infrared. They are used in optical fibre systems. It might be mentioned that, in the days before electronics, visible light was used for communicating by visual signals. Techniques included smoke signals, beacons, semaphore, heliograph and signal flags. All techniques were line of sight transmissions.

Among the many systems using electromagnetic radiation, we list a few examples:

TV remote controllers: The source is an infrared LED radiating pulse coded signals.

Car engine immobilisers: Emit radio signal.

Remote control for toy vehicles and robots uses pulsed infrared or radio. The 315 MHz and 433 MHz frequencies are available for use by hobbyists to use, unlicensed.

Optical mice (for computer control) use LEDs to transmit 1500 images per second of the surface below the mouse. Digital signal processors in the computer interpret the change in the images as the mouse is moved.

RFID radio-frequency tags are passive devices that can be attached to almost anything, and even implanted in the human body. When the tag detects the magnetic field generated by the receiver, it emits an identifying radio signal that the receiver can interpret. In a security system it identifies a wafer card, allowing authorised holders to pass through. In a stock room it identfies how many of each kind of stocks are present and sends this information to a computer for making an inventory, At the entrance to a freeway it can identify each car as it passes by and automatically charge the freeway fee to the owner's account.

The satellites of the global positioning system use the 390−1550 MHz UHF band. This band is also used for communicating with other satellites.

Description	Abbreviation	Frequency Range	Wavelength Range	Propagation Method	Applications
Gamma rays	γ-ray	$> 1 \times 10^{20}$ Hz	$< 1 \times 10^{12}$ m	Penetrative; ionising	Nuclear physics
X-rays	HX and SX	$> 3 \times 10^{16}$ Hz	< 10 nm	Penetrative; ionising	X-ray astronomy
Ultraviolet	UV	750 THz–3×10^{16} Hz	400–10 nm	Ionising	
Visible light		400–750 THz	750–400 nm	Line of sight; optical fibre	Telecomms
Infrared	IR	300 GHz – 400 THz	1 mm–750 nm	Line of sight; optical fibre	Telecomms
Extremely high frequency	EHF	30–300 GHz	10–1 mm	Line of sight	Microwave links, experimental, amateurs
Super high frequency	SHF	3–30 GHz	100–10 mm	Line of sight	Microwave links, radar, amateurs
Ultra high frequency	UHF	300 MHz–3 GHz	10–1 mm	Line of sight Skywaves	TV, telecomms, radar, mobile phones, citizen band
Very high frequency	VHF	30–300 MHz	1 m–10 mm	Line of sight Skywaves	Telecomms, TV, FM radio, amateurs, control
High frequency	HF	3–30 MHz	100–10 mm	Line of sight; sky-waves	Telecomms
Medium frequency	MF	300 kHz–3 MHz	1 km–100 m	Ground waves	Telemetry control, telecomms
Low frequency	LF	30–300 kHz	10–1 km	Ground waves	Telecomms
Very low frequency	VLF	3–30 kHz	100–10 km	Ground waves	Experimental, telecomms
Voice frequency	VF	300 Hz–3 kHz	1000–100 km	Ground waves	AM radio, marine, aircraft
Extremely low frequency	ELF	30–300 Hz	10 000 – 1000 km	Ground waves	

The electromagnetic spectrum. Frequencies are allocated by national and international organisations. Some of the applications are allocated only a small range of frequencies within the bands. For instance, the Global Positioning System uses only two frequencies within the L-Band (390–1550 MHz) which is a sub-band of the UHF band.

BROADCAST RADIO

Many kinds of radio transmitter have an antenna that radiates the waves in all directions with more-or-less equal strength. It is **broadcast** to everyone within range. This is a feature of radio communication that other forms of telecommunication do not normally have. A radio signal spreads through free space. We can communicate with many people at once, as when politicians or disc jockeys speak to their audiences by radio.

Another benefit of radio is that we can broadcast messages to particular individuals without having to know exactly where they are. Ship-to-shore radio, mobile telephones and paging systems rely on radio to convey messages to receivers that are not in a known location.

GROUND WAVES

A signal that is broadcast in all directions can travel from the transmitter to the receiver in two ways. If the frequency is less than 500 kHz, it spreads close to the surface of the Earth as a **ground wave** (see Figs 31.1 and 31.2). It is not able to pass out of the atmosphere because of the ionised layers referred to earlier. These have a lower refractive index than the atmosphere, so the radio waves of lower frequency are reflected back toward the Earth's surface.

Ground waves of lower frequency are more affected by the nature of the surface, particularly when it consists of moist soil or water. The wave fronts nearest the surface are slightly held back so that, instead of being vertical, they are tilted forward. The result of this is that the waves follow round the curvature of the Earth. VLF transmissions can travel for thousands of kilometres in this way and can reach almost any part of the world.

Unfortunately, the surface of the Earth is not completely smooth. Ranges of hills can block the path of ground waves so that the signal does not reach the area just beyond the hills. This limits the range of transmission of ground waves in many terrains. However, if the range has a sharp edge, the waves may be diffracted and then stations immediately behind the hills are able to receive the signals.

SKY WAVES

As frequency increases above 500 kHz the radio waves are able to penetrate the lower ionised layers. As they pass up through the layers the refractive index is decreasing. The effect is similar to that of light passing along a graded index optical fibre. The waves are refracted along a curved path and eventually directed back to Earth again.

These are known as **sky waves**. They reach the Earth at a considerable distance from the transmitter and are important for long-distance communication. The main drawback is that reception of sky waves varies in strength. This is because it is the ultraviolet radiation from the Sun that produces the ionised layers. The layers are not permanent, varying with the time of day, the season, and the number of sunspots.

At a certain distance from the transmitter there may be a point that is too far away for the ground waves to reach, but is too near for the sky waves to reach. The signal is not received at this distance, which is known as the **skip distance**.

Because of the curvature of the Earth and the ionised layers, some of the sky waves skim past the Earth's surface and once more head up toward the ionised layers. Again, they are refracted back toward the surface. In this way, the signal from a sufficiently powerful transmitter may hop around the world.

At frequencies higher than about 25 MHz, radio waves are able to penetrate the ionosphere and pass out into Space. These frequencies are therefore suitable for communication with spacecraft and interplanetary missions.

On the earth's surface, the transmission path for high-frequency signals must be a straight line from transmitter to receiver, otherwise the signals are lost into Space. These are known as **line-of-sight** transmissions. Repeater stations are located on high ground to relay the transmission to areas out of sight of the main transmitter.

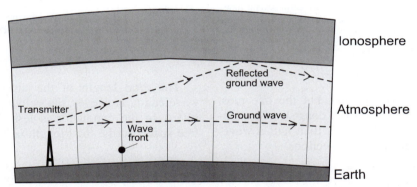

FIGURE 31.1   Ground waves travel in a direction more or less parallel to the earth. Although low frequency ground waves can encircle the Earth, those of higher frequency have relatively short range and are often blocked by high ground. This drawing is not to scale.

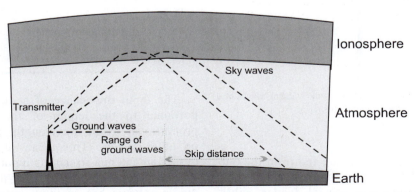

FIGURE 31.2   Radio waves with frequencies between about 500 kHz and 25 MHz are refracted back to Earth in the ionosphere, because of the lower refractive index of the ionised layers. They then become sky waves, which are important for long-distance communication. This drawing is not to scale.

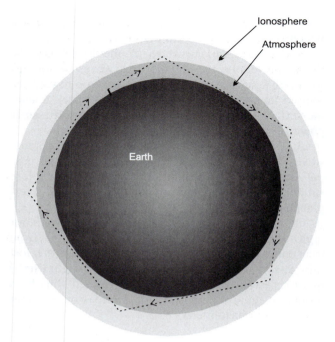

FIGURE 31.3 In good conditions, sky waves may be repeatedly refracted in the ionosphere and circulate round the Earth several times. This produces annoying echoes, repeated at intervals of about 1/7 second. This drawing is not to scale.

ANTENNAS

There are many types of antenna, but the one most often used is the **dipole** (below). It consists of a vertical metal rod or wire, **slightly** less than half a wavelength long, and divided into two sections.

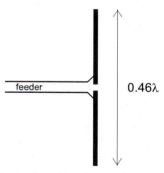

FIGURE 31.4 The motion of the electrons in the standing waves of a half-wave dipole antenna generates radio waves.

The two halves of the dipole are fed from the output of the transmitter, usually by a coaxial cable (75 Ω). An alternative is a special **feeder cable**, which has two stranded copper conductors running parallel with each other and spaced about 9 mm apart.

The halves of the dipole are insulated from each other, so it acts as an open circuit at the end of a transmission line. Standing waves appear in the dipole and in the connecting cable.

An antinode of the voltage waves is located at the ends of the dipole and a node at its centre. The situation in each of the rods is like that in the terminal quarter wavelength of the cable shown in Figure 31.6b.

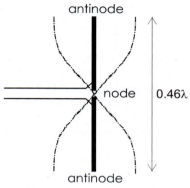

FIGURE 31.5 A standing wave approximately λ/4 long is developed in each half of a dipole, so that the whole dipole is approximately λ/2 long.

The total length of the dipole would be half a wavelength. However, the velocity of electromagnetic waves in the conductor is less than it would be in free space, so the total length is 0.46λ, not 0.5λ.

> **Self Test**
>
> What is the length of a dipole suitable for use with a VHF transmitter operating at 205 MHz.

> **Example**
>
> For a UHF transmitter operating at 554 MHz, the wavelength is $\lambda = 3 \times 10^8/554 \times 10^6 = 0.54$ m. The dipole length is $0.54 \times 0.46 = 25$ cm.

The input impedance of a dipole antenna is proportional to the length of the rod and inversely proportional to the wavelength. It is also dependent on other factors. This means that it is possible to design an antenna so that its input impedance matches the characteristic impedance of the feeder. A typical dipole antenna has an input impedance of 75 Ω and needs a feeder cable of the same impedance for maximum power transfer.

With high-frequency transmissions, the cable is several wavelengths long. A succession of waves enters the cable from the source and these are carried to the far end. What happens there depends on what is connected there between the two conductors:

Matched load circuit (having an impedance equal to Z_0): all of the power of the signal is transferred to the load.

Short circuit: no power is dissipated in this so the power is returned to the cable in the form of waves travelling back toward the source. Where the two conductors are short-circuited, they must be at equal voltage. The amplitude of the wave is zero at this point. Interference between the forward and returning waves sets up a pattern of standing waves as shown in (a). There are alternating nodes (zero amplitude) and antinodes (maximum amplitude). The nodes are spaced $\lambda/2$ apart, with an antinode at the source end (representing the alternating voltage signal supplied by the source) and a node at the far end (because it is short-circuited).

Open circuit: as with a short circuit, no power is dissipated at the far end, so the waves are reflected back along the cable (b). The far end is a point of maximum voltage change, so an antinode occurs at this point. As with a short circuit, there is an antinode at the source end.

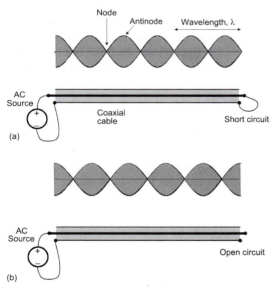

(a)

(b)

FIGURE 31.6 What happens when a signal is sent along a transmission line depends on the loading at the far end. If this is (a) a short circuit or (b) an open circuit, a pattern of standing waves is set up along the line.

Unmatched receiving circuit (having an impedance that is not equal to Z_0): some of the power is transferred to the load, but some is reflected back along the cable. This produces standing waves of smaller amplitude.

The dipole is an example of a **standing wave antenna**, in contrast to some other types, which are **travelling wave antennas**. The formation of standing waves means that the dipole is resonating at the frequency of the carrier wave, and radiating part of the energy fed to it as electromagnetic radiation.

At a receiver, a dipole antenna intercepts arriving electromagnetic radiation, which induces currents in it.

If the frequency is correct, the antenna resonates and standing waves are formed in it.

The antenna receives signals at its resonant frequency much more strongly than those of other frequencies. This helps to make the receiver more **selective**. Although a dipole transmits and receives best at its resonant frequency, it is also able to operate reasonably well at frequencies that are within $\pm 10\%$ of this.

DIRECTIONAL TRANSMISSIONS

The vertical dipole radiates equally strongly in all horizontal directions, so is suitable for broadcasting. But it is wasteful of power to broadcast a signal in all directions when it is intended for only one particular receiver.

The dipole may be made directional by adding further elements to it. Figure 31.7 shows a dipole with two **parasitic elements**, known as the **director** and the **reflector**. These have about the same lengths as the dipole elements. In an antenna being used on a transmitter, the parasitic elements pick up the signal from the dipole, resonate at the signal frequency, and re-radiate the signal.

FIGURE 31.7 Addition of a reflector and director makes the dipole antenna directional and increases its gain.

for growing bacterial cultures) the set point may be fixed. In others it may be altered occasionally (for example, a room heater).

In a **servo system** the set point is often changed and the system is continually adjusting to new values of the set point. Examples are steering systems in vehicles, in large vessels, and in aeroplanes (including flying model planes, Figure 33.7). Other examples are the systems for controlling the position of robot arms.

The systems described in this topic are analogue systems. In Part 3 of the book we describe several other kinds of control system, which are based on digital electronics. The regulator and servo systems described in this topic can all be realised as digital circuits, usually with a microcontroller in charge of the processing.

TEMPERATURE CONTROL

A system which controls an electric heater to hold the temperature of a room at a fixed level is an example of a regulator system. The **set point** can be adjusted from time to time to suit current conditions.

The circuit below is a fully automatic system. It could be used to control the temperature of an oven, a refrigerator, an incubator, a room or a building.

The temperature sensor (R1) is a thermistor. This is part of a potential divider and the feedback signal is the voltages at the junction of R1 and R2. In the diagram, this is labelled t_0. It goes to the inverting input of the operational amplifier IC1, which is connected as a comparator.

FIGURE 33.2 A practical thermostat circuit: the output voltage t_o from the thermistor/resistor chain depends on the resistance of the thermistor, R1. As temperature increases, the resistance of R1 increases and t_0 falls. This is compared with t_s the fixed voltage of the potential divider based on VR1.

Feeding t_o to the inverting input gives it a negative effect — it is negative feedback.

The set point voltage t_S comes from the wiper terminal of VR1, and goes to the non-inverting input of the op amp. This makes the circuit the equivalent of the comparator shown in Figure 33.2.

The op amp is wired as an open-loop amplifier so has very high gain (A). The output of the op amp is:

$$A(t_S - t_o)$$

This is the **error signal**. Because the open-loop gain of the op amp is so high, the error signal swings fully toward the positive or negative rails, turning the transistor switch fully on or fully off, depending on whether t_o is less than or greater than t_S.

FEEDBACK IN A THERMOSTAT

As well as the action of the electronic circuit itself, feedback in a closed-loop system depends on the physical nature of the different parts of the system. Important factors include:

- **The heating element:** A thin wire heats up quickly, so the system responds quickly. We say that it has a rapid **response time**. By contrast a heavy-duty element enclosed in a thick metal sheath heats up slowly.
- It also continues to give off heat for some time *after* it has been switched off. We say that there is **overshoot**.
- **The way heat travels from the heater** to the place to be heated: Parts of a room may be a long way from the heater and are heated only by convection currents. Response time is long. A fan heater has a shorter response time.
- **The response time of the sensor:** A thermistor may be a relatively large disc with long response time, or it may be a minute bead with short response time. The sensor may be in contact with the air or enclosed in a glass bulb.

The effects of these factors produce hysteresis. The heater is switched on at a higher temperature and off at a lower one. It is usually better if the system has hysteresis. It avoids the system being frequently switched on and off when it is operating close to the set point.

In a heating system with hysteresis, the heater goes off when the temperature reaches a given higher level t_U, the **upper threshold**. With the heater off, the temperature falls but the heater does not come on again until the temperature falls below t_L, the **lower threshold**. Then the heater stays on until the temperature reaches the higher threshold. The result is that the temperature is not absolutely constant, but rises and falls over a restricted range.

ROTATIONAL SPEED CONTROL

Varying the current through the coils of a DC permanent-magnet electric motor controls its speed. There are several ways of doing this:

- **Manual control:** A variable resistor is wired in series with the coils.
- **Transistor control:** The speed of the motor is controlled by varying the base current to the transistor. Like the manual control technique, this is open-loop control. It can not guarantee that the speed of the motor is exactly related to the setting of the variable resistor or to the size of the base current.

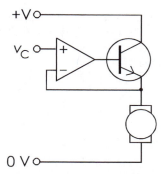

FIGURE 33.4 An op amp with negative feedback. Varying v_C sets the speed of the motor.

FIGURE 33.5 A pulse generator with variable mark-space ratio is used to control the rate of rotation of the motor.

FIGURE 33.3 Changing the control voltage v_C varies the speed of the motor. This is an open loop, so speed depends partly on the mechanical load on the motor.

- The reason for this is that the mechanical load placed on it affects motor speed. This may vary for several reasons, for example: the motor is driving a vehicle which starts to go up a slope; the motor is lifting a load which is suddenly decreased; the motor is affected by uneven friction in the mechanism. A slowly rotating motor may stall completely if its load is suddenly increased.
- **Op amp closed loop control:** The circuit in Figure 33.4 has negative feedback. The op amp monitors the voltage across the motor and acts to keep it constant, equal to the control voltage, v_c. This ensures constant current through the coils, keeping it turning at constant speed in spite of variations in the load.
- **Switched mode control:** Instead of controlling the amount of current supplied to the motor, the current is supplied in pulses of constant amplitude but with varying mark-space ratio.

  The speed of the motor depends on the *average* current being delivered to it. In Figure 33.5, pulses are generated by a timer IC. The timer runs at 50 Hz

or more, producing a stream of pulses. The inertia of the motor keeps it turning at a steady speed during the periods between pulses.

Adjusting VR1 varies the pulse length and thus the mark-space ratio of the output to Q1. With the values given in the diagram, the mark-space ratio ranges from 2 to 50. This gives a 1 to 25 variation in the amount of current supplied to the motor.

This technique gives smooth control of motor speed over a wide range. There is less tendency to stall at low speeds because the pulses switch the transistor fully on and the motor is driven with full power. It is possible for the pulses to be generated by a microprocessor for fully automatic speed control. The main disadvantage of the method is that it is an open loop. The actual speed attained depends on the load.

- **Stepper motor:** A typical stepper motor has four sets of coils, arranged so that the rotor is turned from one position to the next as the coils are energised in a fixed sequence. This is listed in the table following. The sequence then repeats after this. At every step, two coils are on and two are off.

Step no.	Coil 1	Coil 2	Coil 3	Coil 4
0	On	Off	On	Off
1	Off	On	On	Off
2	Off	On	Off	On
3	On	Off	Off	On

To get from step to step, first coils 1 and 2 change state, then 3 and 4 change state, then 1 and 2, and so on. The result of this is to produce clockwise turning of the rotor, 15° at a time. If the sequence is run in reverse, the rotor turns anticlockwise. Six times through the sequence causes the rotor to turn one complete revolution.

At any step, the rotor can be held in a fixed position by halting the sequence. A pulse generator, containing the logic to produce four outputs according to the sequence in the table, controls the motor.

The four outputs are connected to four transistors, which switch the current through each coil on or off. The pulses that switch the transistors are provided either by a special IC or by a programmed microcontroller.

The rate of rotation is controlled by the frequency of pulses generated by the microcontroller, one revolution for every 24 pulses. The microcontroller can be programmed to produce the pulses in the reverse sequence, so reversing the direction of rotation. It can also deliver any required number of pulses to control the angle turned.

The advantage of the stepper motor is that its speed is controllable with the same precision as the system clock. It is not affected by the load on the motor, except perhaps an excessive load, which might completely prevent the motor from turning. The stepper motor may be made to turn 7.5° per step by using a slightly different switching pattern. Motors with a 1.8° step angle are also produced.

POSITION CONTROL

There are two types of position that may be controlled:

- **Linear position:** The location of an object (such as a cutting tool) along a straight path, measured from a point on the path.
- **Angular position:** The direction of a lever-like object (such as an arm of a robot), measured as an angle from some reference direction.

If an object, such as a sliding door, or a part of a machine, is to be moved along a straight path, the motive force may be provided by a solenoid, an electric motor, or a hydraulic cylinder and piston. The rotary motion of the motor may be converted into linear motion by a mechanism such as a worm gear or a rack and pinion. A closed-loop control system can supply power to a solenoid, a motor or the valves of a hydraulic system, but it needs a sensor to detect that the moving object has reached the required position. Sensors used for this purpose include limit switches, magnetic proximity sensors, linear inductive proximity sensors, and various resistive and optical sensors.

FIGURE 33.6 As the object moves, the wiper is moved along the resistive track. The output voltage varies from 0 V up to +V as the object moves from left to right.

A simple resistive linear position sensor is illustrated above. The output of this is proportional to the position of the object. This is compared with v_C, the control voltage, set by the operator or perhaps generated by a computer.

The object is moved along its track by an electric motor and a linear mechanism. The control systems is illustrated below. There are three situations:

- $v_C > v_S$: The output of the comparator is positive. The control circuit supplies current to the motor, so as to increase v_S.
- $v_C < v_S$: The output of the comparator is negative. The control circuit supplies current to the motor in the opposite direction, so as to decrease v_S.
- $v_C = v_S$: The output of the comparator is zero. The control circuit supplies no current to the motor. The object stays in the same position.

This is an example of a **servo system**. It is an on-off system, with the extra feature that the motor is not just switched on and off but may be made to turn in either direction. The control circuit is built from op amps and transistors, or it may be a special IC designed for servo driving. If the object is not at its intended position, the motor rotates one way or the other until $v_C = v_S$.

One of the possible snags of such as system is that the system overshoots. Instead of homing steadily on the intended position, the object is moved *past* that

FIGURE 33.7 A small servomotor of the kind used in flying model aircraft and robots. The 'horns' (white levers) are used for connecting the motor to the mechanisms that it drives.

position. It is then moved toward the position from the opposite side, and may overshoot again. This may continue indefinitely, causing the object to oscillate about the position instead of settling there. This is called **hunting**. To avoid hunting, most servo systems provide a **dead band** on both sides of the intended position. The motor is switched off as soon as the difference between v_C and v_S becomes less than a given small amount. The object stops a little way short of the intended position but the error is small enough to be ignored.

A **servomotor** is designed to move to a given angular position. The motor has three connections to the control circuit. Two of these are the positive and 0 V supply lines. The third connection carries the control signal from the control circuit, which may be a microprocessor.

The rotor of the motor has limited ability to turn. Generally it can turn $60-90°$ on either side of its central position.

The control signal is a series of pulses transmitted at intervals of about 18 ms, or 50 pulses per second.

The angle of turn is controlled by the pulse length:

- 1 ms: turn as far as possible to the left.
- 1.5 ms: turn to central position.
- 2 ms: turn as far as possible to the right.

Intermediate pulse lengths give intermediate positions.

ACTIVITIES — CONTROL SYSTEMS

Set up demonstration controls such as:

- a thermostat.
- an automatic electric fan for cooling.
- motor speed control.
- lamp brightness control.
- stepper motor driver based on a walking ring counter.
- servomotor driver, based on a 555 timer running at 50 Hz, that drives an edge-triggered delay to produce pulses varying in length from 1 ms to 2 ms.

These circuits should be powered from batteries or low-voltage PSUs, not from the mains. Some of these systems could be controlled by a microcontroller.

QUESTIONS ON CONTROL SYSTEMS

1 Draw a block diagram of a switched-mode closed-loop circuit for controlling the speed of an electric motor. Describe the action of each part of the system.
2 Explain how a thermostat operates.
3 What are the advantages of a closed-loop system when compared with an open-loop system? Illustrate your answer with examples.
4 Give examples to explain what is meant by (a) hysteresis, (b) overshoot, and (c) dead band.
5 What is a stepper motor? How can it be used for controlling (a) rotational speed, and (b) angular position?
6 What are the differences between stepper motors and servomotors? For each type, give an example of a control circuit in which you would use it.
7 List and describe in outline the control systems used in three examples of domestic equipment.

The companion site has several animated diagrams of circuits that are used in control systems.

Set up some of these on a breadboard, or build them into a permanent project.

Process Control Systems

This topic looks more closely at the structure and behaviour of control systems, particularly those used in industrial processing.

We are concerned only with closed loop systems as open-loop systems are unable to operate as regulators or as servos.

The simplest closed-loop systems are like the thermostat described in the previous topic. In this the heater is either on or off. This type of system, often known as an 'on-off' or 'bang-bang' system, is relatively easy to implement. It operates well enough in many applications.

Many industrial applications demand more elaborate control systems. There are three main types:

- Proportional (P) systems.
- Proportional-integral (PI) systems.
- Proportional-integral-derivative (PID) systems.

In this topic, the systems operate with analogue voltages. However, the same control functions are possible in digital systems. Signals are digital quantities and many of the circuits are replaced by software programs.

PROPORTIONAL CONTROL

Figure 34.1 shows the main functional blocks of a proportional control system and the analogue voltage signals that flow from one block to another.

The system is shown controlling an industrial processing plant. It might, for example, be a chemical plant in which several solutions are mixed in a vat to react together. But the same principles apply to other systems such as the control system of an underwater vessel used for repairing an oil well platform.

In the figure below the function of comparing set point and output is shown divided into two stages:

- A **subtractor** to measure the difference between the actual output and the set point (or desired output), x.
- An **amplifier** to amplify the difference.

The amplified signal drives the actuators, the motors, heaters, stirrers and other units in the processing plant. There may be further amplification in the plant.

In most controllers the plant introduces a time factor, usually a delay. For example, a controller may fill a vat with a solvent. The system operates a valve and the level of solvent in the vat is measured by a float mechanism.

The valve is controlled by the signal y. The amount of solvent in the vat is detected by the float mechanism. The output from this sensor is used as feedback. The feedback is used to calculate the value of y.

Filling the vat to the set point level is a process that takes time — perhaps several minutes or even hours. If the vat is empty, y is large and the valve is

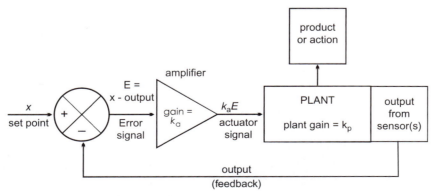

FIGURE 34.1 A signal flow diagram of a proportional control system.

- The heater does not switch off when the set temperature is exceeded.
- Nothing happens when a control button is pressed.

To be able to provide such answers, the engineer must be very clear about what the system should do. Read the User Manual and consult it often.

With some systems it may also be informative to answer this question:

What does it do that it should NOT do?

Typical answers could be:

- There is a smell of scorching.
- The indicator LED flickers irregularly.
- There is a loud buzz from the loudspeaker.
- The servomotor never settles to a fixed position.

Another question that should be asked is:
WHEN does it show the fault?

It may be necessary to run the system for several hours to answer this fully, but typical answers are:

- All the time.
- Not often. It becomes faulty and then recovers.
- Not often, but after the fault develops it stays.
- When it is first switched on, but not later.
- Not when it is first switched on, but starts about 10 minutes later.
- Only when the gain is set high.

INSPECTION

With the fault report in mind, the next stage is to inspect the circuit, looking for obvious faults. These visible faults, if any, should be carefully listed.

Look for wires that have broken away from the circuit board or the terminals of panel-mounted components, signs of mechanical damage to the circuit board and components, or signs or smells of scorching.

At the same time it is important to note what parts of the system can not be inspected and why. In some systems there are complicated components (such as a microcontroller, or a RAM chip) that can not be tested without special equipment. These could be looked at later if the initial inspection fails to deliver any solution to the problem.

Mains Supply

Turn off the mains supply and remove the plug from the socket before inspecting or testing mains-powered equipment.

Unless you are fully experienced you should never test or operate mains-powered equipment with the case open. You should never attempt to test the circuit while it is plugged into a mains socket.

PRELIMINARY DIAGNOSIS

When the inspection is complete it may already be possible to decide what the cause of the fault may be. The next stage is to carry out some elementary practical tests. It is not possible to provide a set of routines here because so much depends on the type of system and the nature of the fault. Some manufacturers provide checklists of faults and remedies applicable to their equipment. If such help is available, it should be followed.

In the absence of advice from the manufacturer, or even when advice is available, it is essential that the person looking for the fault fully *understands* the system and how it works. A schematic diagram is almost essential, preferably marked with typical voltage levels at key test points.

With circuit diagrams and other data to hand, and with a good understanding of the system, it is usually possible to eliminate certain possibilities and to concentrate on locating particular kinds of defect.

Examples

- A total failure of the equipment to do anything at all suggests that there may be a failure of its power supply. With mains equipment, first check that the mains socket provides mains current when switched on (try it with some other piece of equipment plugged in).
- With a battery-powered circuit, check the battery voltage, remembering that the voltage from a nickel-cadmium rechargeable battery drops sharply as it reaches full discharge. It might have been correct yesterday, but is it still delivering full voltage today?
- From the circuit diagram, pick out points in the circuit that should be at full positive supply voltage, and check these with a meter. Such points include the positive supply pins of all integrated circuits and all other pins (such as unused reset pins) that are wired permanently to the positive supply.
- If the supply voltage is low at any of these points, place an ammeter in series with the power supply and the equipment and measure the current it is taking. If the current is excessive, it suggests that there may be a partial or total short-circuit between the positive supply rail and the ground (0 V) rail. This could be due to a defective component.
- Use a continuity tester to check that all points that should be connected directly to the 0 V rail are in fact connected. This includes 0 V pins of all ICs and all other pins that are wired permanently to 0 V.

If a list of expected voltage levels is not available, try to work out what they should be.

Example

If the base-emitter voltage of a BJT is more than about 0.7 V the transistor should be on.

This means that its collector voltage should be relatively low, perhaps not much greater than its base voltage. Conversely, if the base-emitter voltage is less than 0.7 V, the transistor should be off and its collector voltage may be equal to the positive supply voltage.

Capacitors

In many types of circuit, the capacitors retain their charge for a long time, perhaps for several hours, *after* the power supply has been switched off. They can deliver a very unpleasant and possibly lethal shock.

Treat all capacitors with respect and make sure they are discharged before touching them or their terminal wires. Capacitors of large capacity should always be stored with their leads twisted together.

COMMON FAULTS

There are some very basic faults that occur more often than faults of other kinds. It is best to look for these first. Several of them are related to the circuit board and construction techniques. Quality control should eliminate most such faults from commercially-built systems, but self-built systems should be checked *before* power is first switched on:

- There are short-circuits caused by threads of solder bridging the gaps between adjacent tracks on the pcb. A hair-thin thread can result in a fault, so it is *essential* to use a magnifier when inspecting a circuit for this kind of fault.
- There are hairline gaps in the tracks. Breaks in tracks may result from flexing due to mechanical stress in mounting the board. These gaps are difficult to see and it may happen that they only open up after the board has been warmed from some time. Faults of this kind may develop after the circuit has been switched on for several minutes.
- Cold-soldered joints are the result of a lead or the pad (or both) being too cold when the solder was applied. It may also be the result of a greasy or corroded surface. The molten solder does not wet both surfaces properly. Often the solder gathers into a ball on the track or lead and the fact that the joint is dry can be seen under a magnifier. Suspected cold-soldered joints should be resoldered.
- Components may be soldered on to the board with the wrong polarity. This applies particularly to

diodes, LEDs, electrolytic capacitors and tantalum capacitors. It is also possible (and damaging) to insert an IC the wrong way round in its socket.
- It is possible that a wrong IC has been used. Check all type numbers.
- An IC may be inserted in its socket so that one or more pins is bent under the IC and does not enter the socket. This type of fault is hard to see. The only method is to remove the IC from the socket and inspect its pins.
- Resistors and capacitors may be mechanically damaged by rough handling so that leads lose their electrical connection with the device without actually becoming detached. With ICs, excess pressure on the IC may flex the silicon chip inside so that connections are broken.
- Devices such as diodes and transistors may be totally damaged or their characteristics may be altered by excessive heat. This usually happens when the device is being soldered in, but may also be the result of excess current. The resistance of resistors may also be altered by excessive heat.

ANALOGUE CIRCUITS

The items listed above may lead directly to the cause of the fault, which can then be rectified. If this fails, a more detailed inspection is required. In this section we deal with analogue circuits. Suggestions for digital circuits are listed in the next section.

Analogue equipment is checked by further use of a testmeter and with more elaborate equipment such as an oscilloscope. Signal generator and signal tracers are often used with audio equipment. Detailed instructions for using these are found in their Users' Manuals.

The probe of an oscilloscope has high input impedance so the instrument can be used in the same way as a voltmeter for checking voltage levels at key points. In addition, it is possible to detect rapid changes in voltage to discover, for example, than an apparently steady voltage is in fact alternating at a high rate. In other words, the circuit is oscillating. Conversely, we may discover that a circuit which is supposed to be oscillating is static. Clues such as these, combined with the understanding of circuit operation, may point to the cause of the fault. In particular, the switching of transistors may be examined, to see that the base (or gate) voltages go through their intended cycles and that the collector and emitter (or source and drain) voltages respond accordingly. A dual-trace oscilloscope is invaluable for checking transistor operation.

A software fault which causes the microprocessor to perform the wrong action.

briefly state how you would attempt to ...

(a) common-emitter amplifier.
(b) common-drain amplifier.

A signal generator may be used for introducing an audio-frequency signal into an audio system. In a properly functioning system, a tone is then heard

The more detailed checks require the use of special equipment. A **logic probe** is the simplest and most often used of these. It is usually a hand-held

FIGURE 36.2 When the temperature is below the set point, the MOSFET is on and the input to the microcontroller is logic 0. The set point is adjusted by varying R1.

This circuit is used to indicate one of two states: the temperature is either *below* the set point (logic 0) or it is *above* it (logic 1). If the output of the circuit is likely to hover at the midway point, the output is fed to a Schmitt trigger circuit before being sent to the controller. A Schmitt trigger is also used to eliminate the effects of noise in the sensor circuit. The transistor switches of Topic 2 can all be adapted to provide 1-bit input.

> **Set Point**
>
> In this case the temperature at which the transistor is to switch on or off.

A 1-bit input is used to measure linear position in a **bar encoder**. A device such as this relies on the programming of the controller to interpret the input signal.

FIGURE 36.3 A bar encoder has a plate with alternate transparent and opaque bars. At (a) a single LED allows the distance and speed to be measured. At (b) a pair of LEDs allow direction to be measured as well.

The barred plate moves between an LED (usually infrared) and a photodiode. The photodiode is interfaced to the controller. In (a) the controller alternately

receives '0' and '1' as the plate moves. If the controller is programmed to count the 0 to 1 transitions, the position of the plate (and of the machine part to which is is coupled) is known. By timing the transitions the controller can also calculate the speed of motion.

With two LEDs and photodiodes and a 2-bit input, as in diagram (b), the direction of motion can be read as well. The figure below shows the signals.

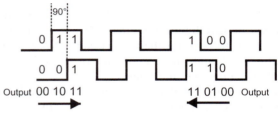

FIGURE 36.4 The output of a 2-bit bar encoder consists of two pulsed signals, one shifted 90° with respect to the other.

When the plate moves to the right, input 2 (upper plot) changes state before input 1 (lower plot). The change from 00 to 11 passes through 10. Conversely, as the plate moves to the left, input 1 changes before input 2. The change from 00 to 11 passes through 01. The controller is programmed to read these changes, so detecting the direction of motion.

The same principle is used to read the rotary motion of a shaft. The shaft bears a circular plate marked with radial bars. The controller can detect the angle turned, the rate of rotation and (with 2 diodes) the direction of rotation.

MULTI-BIT INPUT

An alternative device for position sensing is shown in the figure below. The diagram shows only 4-bit input for simplicity, but input may be 8-bit or more to give greater precision.

FIGURE 36.5 When the encoder is at position 7, as in the drawing, the code read by the optical system is 0100. If the encoder moves to position 8, only one digit, D3 changes and the code becomes 1100.

The plate is transparent but marked with an opaque pattern. There are four LEDs on one side of the plate and four photodiodes on the other side. The pattern is not based on the binary number system but on a sequence of codes known as the **Gray Code**.

The principle of the Gray code is that only one digit changes at a time. As seen from the diagram, only one digit changes as the encoder moves from position 7 to position 8. If the encoder were marked with the binary equivalents, the change would be from 0111 to 1000 — *all* the digits would change. The difficulty is that, owing to slight errors in the alignment of the LEDs and photodiodes, they would not all change at *exactly* the same instant. Going from 0111 to 1000, they would probably change one at a time. We might get a sequence such as 0111, 0101, 0100, 0000, and 1000. This reads as 7, 5, 4, 0, 8. Using the Gray Code makes this type of error impossible.

The Gray Code is also used for shaft encoding, using a circular plate marked in segments.

If a sensor has analogue output, an ADC is often used to convert this to multi-bit input. Each output of the converter is connected to a single data line. One or two output lines from the controller may be used to instruct the ADC when to begin a conversion and to latch the converted value. There may also be a connection from an 'end of conversion' output of the ADC to tell the controller that conversion is completed and that the converted data is on the lines waiting to be read. This is another example of **handshaking**.

ISOLATED INPUTS

In industrial applications it is generally advisable to isolate the controller from external circuits. The voltages used in industrial plant are often higher than those used for powering a controller. Also heavy current switching in industrial circuits produces large voltage spikes that must not be allowed to reach the controller.

The input terminals of programmable logic controllers (PLCs, See 'Programmable Logic Controllers', Topic 37) are isolated from the external circuits by means of a circuit like that shown in Figure 36.6.

The signal from the plant is fed through a dropper resistor and indicator LED to an opto-isolator. When the input is at logic high (say 24 V) the indicator LED is on, and so is the LED in the opto-isolator. The light from this switches on the transistor. The voltage at its collector falls to logic low. This is inverted by the NOT buffer gate, giving a logic high that is sent to the controller.

FIGURE 36.6 One of the 8 or more channels of a typical DC digital input card of a PLC. The supply from the plant may be at a relatively high voltage, such as 24 V, often unregulated. That used by the controller is often at 3 V or 5 V and is regulated.

PLC input cards are also made that operate with an AC input from the plant. An AC voltage is equivalent to logic 1 and zero voltage to logic 0. The AC voltage is passed through an indicator lamp, then rectified by a diode bridge. The resulting DC signal goes to an opto-isolator as in the figure above.

ONE-BIT OUTPUT

Low-current devices, such as an LED, can be driven directly from an output pin of a controller. A series resistor is needed to limit the current to a safe level.

FIGURE 36.7 Driving an indicator LED from an output of a controller.

Other devices can be driven by using a transistor switch to turn them on or off. A circuit is on shown in Figure 2.18, p. 19. Devices that can be controlled in this way include lamps, solenoids, motors, relays, and speakers.

The circuit shown in Figure 33.5 shows a switched mode motor speed control. In this, the transistor is switched by the output from a 7555 astable. Instead of this, it could be switched by the pulsed output from a controller. The controller is programmed to make one of its output pins go alternately high and low. The program can vary the mark-space ratio and

Processing

Personal computers, either the desktop or the laptop versions, are probably the kind of microelectronic system most familiar to most people. They are known as **microcomputers**. There are more complicated systems, such as the **workstations** used by geologists, for example, to analyse seismic data when prospecting for oil. These come somewhere between microcomputers and the **minicomputers** used by medium-scale businesses. The most complex systems are the **mainframe** computers used by the largest businesses such as banks and airlines, or by researchers in highly technological applications such as aircraft design. But the distinctions between types of computers based on size are rapidly becoming out of date. The present day personal computer (PC) has far more power than the mainframes of, say, a decade ago. Also, a number of smaller computers can be networked to work together with the effective power of a **super-computer**.

At the other extreme to the mainframe is the 'computer on a chip', or **microcontroller**. This is a single IC capable of performing all the functions of a computer but lacking some of the connections to the outside world. It may have no monitor, or only a 16-key keypad instead of a full typewriter-style keyboard. Often it has a fixed program built into it, to perform one particular task. Such microelectronic systems are found in hundreds of different devices, from microwave ovens and smart cards to washing machines and video games machines.

The important thing about all these microelectronic systems is that they all have very similar structure (architecture) and all work in much the same way. Some may lack certain stages and others may have specialised additions, but basically they are all very similar. In the descriptions that follow we refer mainly to microcontrollers, the most widely used of the microelectronic systems.

FROM BITS TO TERABYTES

Some of the terms mentioned in this section have been defined earlier in the book, but these are the fundamental units of data for microelectronic systems so they are repeated here, from the smallest to the currently largest:

- **Bit:** the unit of information. Takes one of two binary values, 0 or 1. In logical terms, these may mean TRUE and FALSE. In positive logic, 1 is TRUE and 0 is FALSE, and this is the system most commonly used. In electronic terms, 0 means low voltage (usually very close to 0 V) and 1 means high voltage (usually close to the supply voltage, which is +5 V in TTL systems).
- **Nybble:** The upper or lower four bits of a byte.
- **Byte:** A group of eight bits. Bytes are used in very many microelectronic systems as the basic unit of data.
- **Word:** The unit of data used in a particular system. In an 8-bit system the word is 8 bits long, so a word is a byte. In a 16-bit system, a word is 16 bits long, or two bytes. In a 32-bit system a word is four bytes.
- **Kilobyte (kB):** According to the metric system, this would be 1000 bytes but in digital electronics this term means 1024 bytes, or 2^{10} bytes. This term is often used when describing the data storage capacity of small devices such as the memory of microcontrollers, and the lengths of typical data files.
- **Megabyte (MB):** This is 1024 kilobytes, or 2^{20} bytes, or 1 048 576 bytes. It is *just over* a million bytes. This unit is often used in describing the data storage capacity of devices such as floppy disks (1.44 MB), memory chips, CD-ROMs (650 MB or more), and long data files (in particular those containing photographs with over 16 million different colours).
- **Gigabyte (GB):** This is 1024 megabytes or 2^{30} bytes. It is just over a thousand million bytes. The capacity of the hard disk drives of PCs is reckoned in gigabytes. DVDs store 4.7 GB. This is the largest unit in common use at present but a professional photographer, for example, may use several terabytes (= 1000 GB) for storing a collection of high-resolution digital images.

BUSSES

The block diagram below shows the main features of a microcomputer. It consists of a number of units connected to each other by busses. A bus may be serial or parallel:

- **Serial bus:** A single conductor along which the bits are signalled one at a time by making the line voltage high or low.
- **Parallel bus:** A set of conductors, often 8 or 16 running side by side. At any given moment the busses each carry one bit of a word, so the bus is able to carry a whole byte or double-byte or more of data.

A serial bus is simpler to wire up or to etch on a printed circuit board, but a parallel bus transfers data much more quickly.

Most busses in a microcomputer are parallel busses and are divided into three types:

Data bus: Used for carrying data between different parts of the system. The data may be numerical values, coded text or coded instructions.

Address bus: Used by the microprocessor for carrying addresses of devices (such as the hard disk drive) or locations in memory. When the microprocessor calls a device into action (to receive data, for example) it places the address of that device on the bus. The addressed device then registers the data or instruction that is on the data bus at that time.

Control bus: Used by the microprocessor for controlling the other devices in the system.

Only the microprocessor can place signals on the control and address busses, but the data bus can have data put on to it from several sources. These include the microprocessor, the memories, and the keyboard. In such a system there may be times when several devices may be trying to put data on the bus at the same time. This is referred to as **bus contention** and must not be allowed to happen. To prevent bus contention, each device that may need to put data on the bus has **three-state outputs**.

These outputs are under the control of the microprocessor (through the control bus). At any given instant, a three-state output may be in *one* of the following three states:

- Enabled, with low output.
- Enabled, with high output.
- Disabled, with high-impedance output; in effect the output is disconnected from the bus so that it can put no data on it.

The microprocessor allows only *one* device to be connected at any one time. Only that device is able to place data on the bus.

ELEMENTS OF A MICROCOMPUTER SYSTEM

The devices that make up a typical microcomputer system are:

- **Microprocessor:** This is a highly complicated VLSI device and is the heart of the system. Its function is to receive data from certain parts of the system, act upon it according to instructions presented to it in a program, and output the results. It is described in more detail later.

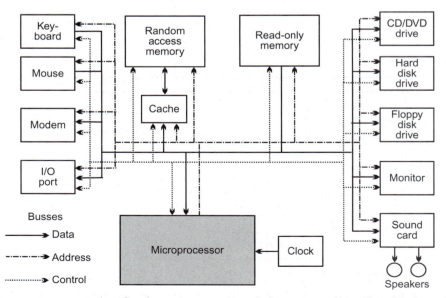

FIGURE 37.1   A microcomputer system consists of a microprocessor, a system clock, memory, and a number of devices to interface the system to the outside world. The parts of the system communicate with each other by placing signals on the busses.

- **Clock:** This is a square-wave oscillator that causes the microprocessor to go through one cycle of operation for each pulse. The clock is usually a crystal oscillator and operates at 2 GHz or more.

  Read-only memory (ROM): This is memory that has permanent content, programmed into it during the manufacture of the system. It is used to store programs that instruct the microprocessor to start up the system when it is first switched on. In a micro-controller system intended for a particular purpose (for example, to control a dishwasher) the ROM contains a number of programs and also stored data. Certain types of ROM can be repeatedly programmed, erased and reprogrammed electronically. These **electrically eraseable ROMs**, or **EEROMs**, are commonly incorporated into microcontrollers.

- **Random-access memory (RAM):** This is a temporary data store, used for storing the whole or parts of the program that is currently being run, tables of data, text files, and any other information which needs to be registered temporarily. Typical ROM size is 512 MB.

- **Cache memory:** Certain systems, particularly those required to run at high speed, include a small amount of special RAM. This consists of chips that operate at very high speed. It would be too expensive for the whole of RAM to be built from such chips. There may also be a small cache memory built into the microprocessor.

Data that is often required is stored in the cache so that it is quickly accessible. Usually the cache is operated so that data on its way to or from RAM is held there for a while. This data is later overwritten. In some systems it is overwritten if it is infrequently used. In other systems, the overwritten data is that which has been there for the longest time. Either way, the data held in the cache is the data most likely to be needed in a hurry.

- **Input/output devices:** To the left and right of the diagram in Figure 37.1 are the units that provide input and/or output to the system. These are often referred to as **peripherals**.

  - **Input** devices include a keyboard, a mouse, and a joystick.
  - **Output** devices include a monitor, a sound card with twin speakers, and a printer.
  - **Input and output** functions are performed by some peripherals. These include a hard disk drive (around 180 GB of storage), a floppy disk drive, a CD/DVD read/write drive, and a telephone modem.
  - **External peripherals** such as scanners and video cameras are usually connected to the system through USB (Universal Serial Bus) sockets. USB sockets also accept flash memory cards for off-line data storage. There may be special sockets for memory cards, such as those from digital cameras.

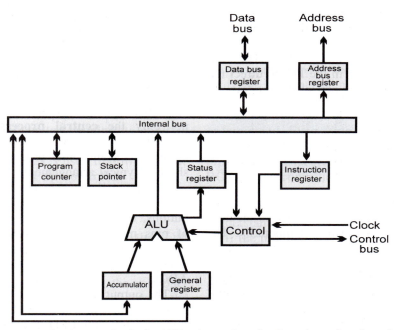

FIGURE 37.2   A microprocessor contains the control unit, the ALU and a number of registers (more than shown here) all connected by one or more internal busses, and all on one silicon chip.

FIGURE 37.6 A screen shot showing a program being written to control a security system. The program is being written in assembler (see 'Programming an Assembler', Topic 39) in the top left window. Other windows show the state of the registers in the processor, the 32 data registers and the input and output registers. The program is run and tested on the computer, using this software to simulate the processor. When ready, it is downloaded into the controller.

A very wide range of controllers is available from Atmel, Microchip and several other manufacturers. When selecting a controller for use in a given project there are a number of features to be considered:

- **Size of program memory:** 1 kB or 2 kB is enough for controlling the simpler systems.
- **Size of RAM:** 64 bytes to 128 bytes is typical.
- **Number of data registers:** these are used for storing and manipulating data.
- **Clock rate:** most controllers are easily fast enough for most applications.
- **Supply voltage:** 2 V to 15 V; most controllers can be battery-powered.
- **Number of input/output pins:** from 6 to 20 or more.
- **Instruction set:** number of instructions. A few instructions (as few as 33 in some cases) make it easier to learn to program the controller.
- **Special on-chip units** (if needed): may include analogue comparator (1 or more), ADCs (1 or more), DACs, op amps, voltage reference, and display driver.

PROGRAMMABLE LOGIC CONTROLLERS

PLC systems are built up from a number of ready-made units that can be mounted on a rack and connected together in various ways. These allow great flexibility in building a system for a particular application.

FIGURE 37.7 The distinguishing feature of a PLC system is the large number of inputs and outputs.

A **programmable logic controller** is a simple computer-like device especially designed for industrial control systems. PLCs are often used for other purposes too, such as controlling the automatic microwave systems that open and close shop doors. A typical PLC is based on a specialised microprocessor.

FIGURE 37.8 The door of this PLC cabinet is open to reveal three PLCs on the right, with numerous connections to the sensors and actuators of the colour-printing press that they control. Relays on the left are used for switching.

Hex

The hexadecimal numbering system.

Example

A system has 11 address lines. The lowest possible address is 000 0000 0000 in binary. All lines of the address bus carry a low voltage. The highest possible address is 111 1111 1111 in binary. All lines carry a high voltage. This is the equivalent of addresses from 0 to 2047 in decimal or from 0 to 7FFF in hex. It is a 2-kilobyte system.

Note

If you find these numbers difficult to understand, turn to the section on 'Number Systems' in Supplement A.

There is a terminal which usually has a few keys for operating the PLC and an LED or LCD display to show the current status of the system. The terminal also contains the program and data memory. The control sequences and logic are set up on a PC or a special programming unit which is not part of the PLC. When the program has been written it is downloaded into the memory of the PLC, which then runs independently.

The most noticeable feature of a PLC is the large number of inputs and outputs. Each input and output 'card' communicates with the outside world through a number of ports, usually 8 or 16 ports for each card. There may be several racks of cards, because controlling a large industrial unit such as a production line or a printing press involves numerous sensory inputs and a corresponding number of outputs. However, for elementary tasks there are smaller PLCs that have only about 6 inputs and outputs, each complete with interface circuits and housed as a single unit.

ADDRESSING

It is important for the processor to be able to communicate with any other part of the system. This is done by giving every part of the system an **address**. The address is in binary, a set of low and high voltage levels on the lines of the address bus, but we nearly always write it in hex to make it easier to read and understand. The address bus has a line for every binary digit of the address. In a microcontroller system the address bus is internal, that is, it is on the controller chip.

Sixteen address lines is a common number for microprocessor systems, but some have 20 address lines (1 MB of addresses) or more. Microcontroller systems need fewer lines, and the address bus is internal, on the chip.

The way that the memory space of a system is allocated to different functions is set out as a **memory map**. A typical memory map of a 64 kB computer system is shown in Figure 37.9. Different parts of the system are allocated their own addresses within this memory space. At the lowest addresses in memory there is a ROM chip, containing instructions to the microprocessor on what to do when the system is being started up. The microprocessor has 0000 in its program register when it is first powered up so it always begins at address 0000 and goes there first to find out what to do. There it is given instructions to jump to some other address, perhaps in another ROM, where it finds instructions for starting up the system.

The microprocessor may also go to this low memory when it has received an interrupt (see later), and is then made to jump to one of several interrupt service routines.

The next area of memory may be used for devices such as sound cards and other peripherals. Often these occupy a small range of addresses and the card has logical circuits to decode addresses within its range and to take action accordingly. In general terms, this is referred to as the input/output area of memory space. The devices with addresses within this range are not memory, but they are **memory-mapped**. That is, they are accessed by the microprocessor by using addresses, just as if they were actual memory.

FIGURE 37.9 The amount of space in a memory map depends on the number of address lines that the system has. The computer system shown here has 16 address lines. The memory space is allocated to RAM, ROM and all the input and output devices of the system. Not all of the memory space is necessarily taken up; there are ranges of spare addresses where future additions to the system may be located. The numbers on the right are starting addresses of blocks of memory, the H indicating that they are written in hex.

RAM for holding programs and data takes up the main part of the address space. Some addresses in RAM are reserved for temporary storage of data. One example is the stack, which is discussed later.

The memory map of a microcontroller has fewer addresses but is similar in principle. There is a program area of a few kilobytes, but this is ROM, not RAM. In the controllers used in program development this ROM is electrically eraseable, as already explained. However, once a design has been perfected and is to be mass-produced, a version of the controller with a **one-time programmable** ROM is used. The program is burned permanently into this and can not be altered.

The controller also has a small area of RAM for the data area and for the input/output ports.

The memory map of the processor unit of a PLC is organised in a similar way, with EEROM for the program area, and RAM for a large input area, a large output area and a data storage area.

PROCESSING

Having looked at the main devices forming a microelectronic system, we will now see how they operate.

This description is based on the operation of a typical controller, such as the PIC or Atmel ICs.

- A controller can do nothing without instructions. The instructions are stored in the program memory as a sequence of operations codes, or **op codes**. The central processing unit (**CPU**) of the controller reads and obeys this sequence of instructions one at a time. The stages of the operation are:
- the CPU takes the address stored in the program counter (**PC**),
- it puts that address on the address bus,
- the addressed unit of memory puts its stored data on the data bus,
- the CPU loads the data into its instruction register,
- The CPU does what the instruction says, usually with the help of its arithmetic-logic unit (**ALU**) and the accumulator.

This process is known as the **fetch-execute cycle**. It is timed by the system clock. The program counter is incremented by 1 every time an instruction has been executed. So the controller goes to the next address in memory to fetch the next instruction. In this way the controller works its way through the program, instruction by instruction, from beginning to end.

To ensure that the controller always starts at the beginning, the PC is reset to 0000H when the power is first switched on or the reset button is pressed. As we shall see, there may be instructions that tell the controller to jump to other parts of the program instead of steadily working its way through it. In such cases the PC is loaded with the address to which the controller is to jump.

Once an instruction is in the instruction register it may be executed straight away or its execution may require further fetch-execute cycles. The CPU may need additional data to work on. However, most controllers execute the program in single cycles.

As an example of a fetch-execute cycle, we will see how the PIC16F690 increments a value held in one of its file registers. The file registers are a set of 256 data registers, each able to hold 1 byte. The accumulator of the CPU, called the working register (W) in this controller, is also 1 byte wide.

In this controller, instructions are not sent along the data bus, but along a special instruction bus, which is 14 lines wide. This allows the instruction register to have 14 bits. Fourteen-bit instructions include all the data needed for their execution, so they are executed in a single cycle. An example of a 14-bit instruction is:

00 1010 0100101

What actually happens is that the lines of the instruction bus are set to 0 V where there is a '0', and to the supply voltage (say, 6 V) where there is a '1'. This particular combination of lows and highs causes the CPU to perform a particular operation. Let us analyse the opcode to see what the operation is.

The opcode can be considered in three parts:

- The first 6 bits are the instruction: '001010' means 'increment a file register'.
- The next bit is the destination bit. If it is '0', place the result in the accumulator. If it is '1', place the result in the file register, replacing what was there before. In this example, the bit is '1', so the result of incrementing goes in the file register.
- The last 7 bits are the address of the file register to be operated on. '0100101' is the binary equivalent of 37 in decimal.

The outcome of this fetch-execute cycle is that whatever value is stored in file register 37 has 1 added to it, and the new value is placed in that register.

Many operations, including incrementing, affect one or more bits (often called flags) in the status register. Incrementing may affect the zero flag, Z. If the value in register 37 happens to be -1, incrementing it results in zero. In this case the Z flag is set to '1'. In all other cases it is set to '0'. Subsequent instructions may read the Z flag and take one of two actions, depending on its value.

As another example, take the following:

$$01\ 0010\ 101\ 0\ 0101$$

This consists of three parts:

The first 4 bits, '0100', give the instruction, which is to clear one particular bit in a file register.

The next 3 bits give the number of the bit to clear. In this example '101' means bit 5.

The last 7 bits give the register number, '0100101', which is 37.

The outcome of this instruction is to clear bit 5 of register 37 to zero. For example, if the register holds 1011 1101 before the operation, it holds 1001 1101 afterward. This operation does not affect the status flags.

The two instructions outlined above are typical. They illustrate the fact that a controller operates in very small steps. A programmer has to break down what may be a very complex operation into a series of very many, very small operations.

The reason that the controller is able to perform complex tasks is that each operation takes a very small length of time to perform. If the processor is running at 20 MHz, a single cycle takes only 50 ns.

Another point illustrated by these examples is that writing a program by setting out the opcodes in binary is a slow and onerous task. There are many opportunities for making mistakes. This is why we normally use special software to help us write and develop programs.

EXTENSION BOX 23 Watchdog Timer

In spite of all the precautions, it may still happen that a bit in memory or in one of the registers of the CPU may become altered by electromagnetic means or by a voltage spike on the supply. Substituting a 0 for a 1 or 1 for a 0 can provide the CPU with the wrong op code, or false data. Worse still, this may make it jump from its present place in the program to some unpredictable place in another part of RAM, where there may be no program. The result may be errors in output or the program may crash completely.

The watchdog timer is a technique for avoiding such dangers. It keeps a watch on the CPU, making sure that it is operating correctly. The timer can be an ordinary 7555 timer circuit. More often, it is built into the controller and runs automatically.

The CPU is programmed to trigger the timer at frequent intervals, perhaps every millisecond. The period of the timer is a little longer than the interval at which it is triggered. When the timer has been triggered, its output goes high and it should be triggered again before the period is over.

The output of the timer is connected to the reset input of the controller and, as long as the timer holds this high, the controller runs normally. If a fault develops and the CPU is no longer following the program correctly, it no longer triggers the timer.

After a millisecond or so, the timer output falls to logic low. This automatically resets the controller, which jumps back to the beginning of the program and starts again.

ACTIVITY — CONTROLLER SYSTEMS

The topics in this topic are best studied by using a microcontroller development system. Use the handbook to learn about the features of the controller that you are using. The handbook will probably contain many example programs for you to run. You may prefer to postpone this activity until after you have studied one of the program languages in the next topic.

QUESTIONS ON PROCESSING

1 Name four kinds of microelectronic system; state who uses them and what they are used for.
2 Explain what bit, nybble, byte and word mean.
3 Describe the three types of bus found in a microelectronic system and the type of information they carry.
4 What is meant by *bus contention* and how may it be avoided?
5 List the differences in properties and uses between RAM and ROM.
6 Draw a diagram of a typical microcomputer system and explain briefly what each part does.
7 Draw a diagram of a typical microcontroller system. In what ways does it differ from a microcomputer system?
8 Draw a diagram of a typical micro controller and explain briefly what each part does.
9 Describe what happens during a fetch-execute cycle of a microcontroller.
10 What is an opcode? Give an example of an opcode and explain how a controller responds when the opcode is loaded into its instruction register.

MULTIPLE CHOICE QUESTIONS

1 The unit of information is a:
   A byte
   B bit
   C nybble
   D word.
2 The number of bits in a byte is:
   A 4
   B 8
   C 16
   D 32.
3 A kilobyte is:
   A 1000 bytes
   B 1024 bits
   C 1024 bytes
   D 1000 bits.
4 The control unit of a microprocessor:
   A reads data from the data bus
   B tells the ALU what to do
   C controls the rate of the clock
   D writes addresses on the address bus.
5 If the result of an operation has the value 0, the zero flag:
   A becomes '1'.
   B does not change.
   C becomes '0'.
   D changes state.
6 The register in which the controller performs calculations is called the:
   A status register
   B accumulator
   C program counter
   D central processing unit.
7 The code used by a microprocessor when first switched on is stored:
   A in cache memory
   B in ROM
   C in RAM
   D on the hard disk drive.
8 The PIC16F690 controller has opcodes that are 14 bits wide. The first six bits of the codes for file register operations tell the controller:
   A what value is to be added to the value already in the working register.
   B the address of the register that the controller is to operate on.
   C which operation the controller is to perform.
   D the program address to which the controller is to branch.
9 If the result of certain operations is zero:
   A the program counter is reset to zero.
   B a bit in the status register is reset to 0.
   C the working register is cleared.
   D a bit in the status register is set to 1.

Programming

In Topic 37 we saw that a controller is made to perform its tasks by reading and obeying a sequence of instructions held in its program memory. The instructions are stored as a binary code, a pattern of 0's and 1's. There is a different code, or **opcode**, for each operation that the controller can perform. This sequence of opcode is known as **machine code**.

The code in which a program is written is known as the **source code**. For a program written in machine code, the source code is machine code. However, most people find it difficult to use machine code as the source code. Instead, they use an easier source code.

One of the simplest source codes is **assembler**. Assembler software is available for running on a PC or other computer. Instead of using binary codes for the opcodes, the programmer keys in instructions in abbreviated form. The software turns these into machine code, which is then downloaded into the controller.

Even easier to use is a high-level language, such as **BASIC**. This name is short for **B**eginner's **A**ll-purpose **S**ymbolic **I**nstruction **C**ode. The user types in the instructions in a form that is very similar to ordinary English.

BASIC is easy to learn, and the programs are easily understood. When used in conjunction with a programming board, the BASIC program is written and tested on a computer. Then the software compiles the BASIC program into its machine code equivalent ready for downloading into the controller. Compilers are also available for other high-level languages, such as C.

Programmable logic controllers (PLCs) are often programmed by using a system called **ladder logic**. The reason for the name becomes obvious when you look at a ladder logic program. Ladder logic is intended for use in industrial control systems. It is easy to learn as it has a small set of instructions and a graphical way of setting out the program. It is designed for use by engineers, rather than by computer programmers.

Before writing any program it is best to set out a **flowchart**. It shows the routes by which the controller runs through the program. In the next section there are flowcharts illustrating most of the actions of which a controller is capable. In Topic 39 we turn

some of these into programs written in machine code, assembler, BASIC and ladder logic.

FLOWCHARTS

A flowchart consists of a number of boxes, representing the stages of an operation. The boxes are joined by arrows to show the sequence of the stages. The shapes of the boxes indicate the nature of the stages.

The routines described in this section are divided into seven main sections: initialisation, arithmetical and logical routines, timing routines, data handling routines, input routines, and output routines. Because they are arranged by function, a few of the more complex routines come early in the list. Skip over these the first time through.

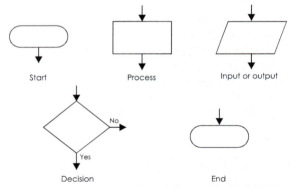

FIGURE 38.1 Symbols used in flowcharts. All programs begin with a 'Start' box. Many have no 'End' box.

In the flowcharts, any numerical values are given an appropriate name, written in italics. Examples are *num1*, *num2*, and *sum* used in the first flowchart. These represent two numbers and their sum.

We shall assume that all variables are integers, with values between 0 and 255. This allows them to be coded as single bytes. Working with larger values or with decimal fractions makes the programming much more complicated. In any case, integers between 0 and 255 cover most of the requirements for programming control systems.

In the flowcharts, which involve the working of the CPU itself, we use symbol A to represent the accumulator. A left pointing arrow ← indicates copying data from one register to another. For example:

$$A \leftarrow num1$$

This means that the data held in the register which has previously been defined as *num1* is copied to the accumulator. This leaves the contents of *num1* unchanged. The left pointing arrow is also used for assignment, for example:

$$sum \leftarrow A + sum$$

This means that the content of the *sum* register is added to the content of the accumulator and the result placed in the *sum* register. The value held in *sum* is also in the accumulator.

INITIALISATION

It is usual for an assembler or compiler program to begin with a sequence of instructions to set up the controller for running the program. The flowchart on the right is an example.

The first thing is to declare the type of controller which is intended to run the program. This information is essential for the assembler or compiler to be able to use opcodes that the controller can understand and act on.

Usually, certain features of the controller need to be configured for the whole of the program. We might need to declare the type of system clock (internal or external) crystal or resistor-capacitor timing, and so on).

Another feature that may need to be set up is: whether the watchdog timer is to be enabled. In the PIC controller this and other settings are coded into a configuration word, which comes next in the program.

Assemblers operate by referring to data stored at numbered addresses in memory. But names are more memorable than numbers so the next thing on the program is a list of the names and the addresses they correspond to. Some of the names may be pre-defined registers such as the status register, the program counter and the input and output ports. Others may be the names of user registers that we allot to important program varibles.

These names, or labels, are for our convenience and are used by the assembler, but the controller itself does not know anything about them. The controller works solely by numerical memory addresses.

The next items may generally be attended to in any order. Usually a port is a byte wide and can be

FIGURE 38.2

set to to input or output bytes of data. Its bits can also be set individually. This is only the inital setting — it can be changed later. If the controller has the ability of accept analogue input instead of digital, analogue inputs can be defined next.

Usually it is safer to clear the port registers at this stage, so that external devices are not accidentally switched on. But other initial values can be sent to output ports or bit, if required.

There may be several program lines devoted to setting up on-chip devices such as timers and analogue-digital converters.

There may also be instructions to enable certain functions, such as interrupts.

Up to this point, the controller has done nothing. The assembler or other programming software has registered information (about the controller and the labels) within the memory of the PC, and set various registers in the controller. All is ready now for the program to be run. The following sections outline some of the things the controller can be instructed to do.

ARITHMETICAL AND LOGICAL ROUTINES

1) **Add two bytes** stored in two memory locations and place their sum in a third location in memory. This is a task to be programmed in machine code and in assembler. It is trivial in BASIC. It illustrates once again the small steps by which the activities of the controller proceed.

   In the flowchart below, the registers used for *num1*, *num2* and *sum*, have already been defined earlier in the program. Also, the two values to be added are already stored in memory at these locations. The flowchart shows how these are added. It is assumed that adding can be done only in the accumulator.

2) **Repeat an action** a specified number of times. The number of times, *n*, is stored in a register labelled *count*.

   The structure of this program is a **loop** (Figure 38.4). It holds a routine (not specified here) which is to be repeated *n* times. The routine might be simple, such as flashing an LED on and off. Or it might be a complex one, such as the spraying of paint on to a car body in the assembly line.

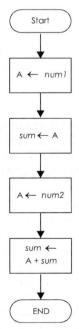

FIGURE 38.3 Adding together two numbers, *num1* and *num2*, to give *sum. Num1, num2* and *sum* are the labels allocated to three bytes of memory. It is assumed that *sum* will always be a positive integer, not greater that 255, as it is held only as a single byte.

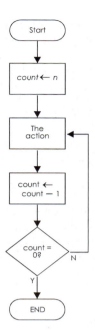

FIGURE 38.4

The routine in the loop could be a simple addition, such as adding 3 to the accumulator. If A starts by holding 0, it finishes by holding 3*n*.

This makes a simple routine for **multiplication**, useful because controllers can add and subtract, but not directly multiply.

3) **Conditional jump** to another part of the program. In the flowchart below, the controller is able to read the room temperature because it receives input from a sensor through a DAC.

   The program has the set point written into it and, at the condition box, the controller determines if the

FIGURE 38.5 The diamond-shaped **condition box** is where the controller is made to take a certain action depending on whether the temperature is above the set point or not.

temperature is above that set point or not. If it is above (Y), the program proceeds to an output box, with an instruction to turn on a fan. If it is not above (N), the controller switches on a heater.

This flowchart is not a complete thermostat, as it takes a once-for-all decision, but it can be easily extended for complete control.

4) Logical AND is implemented by two or more decision boxes. They are connected so that the decision must be Y for both (or all) to allow the controller to get to the action.

The flowchart (see below) is used for an outdoor security floodlamp. The lamp is to be turned on only if it is dark AND the motion of a person has been detected. The lamp does not come on in daylight, even if a person is detected. It does not come on in darkness if no motion is detected.

The system has two sensors. One is a light dependent resistor with a potential divider circuit that gives a high output (1) in darkness and a low output (0) in the light. The other sensor is a passive infrared sensor that detects changes in the amount of infrared reaching it. It is able to detect a moving person at a distance of several metres. It produces a high output when it detects a moving person.

The lamp is turned on only if both decision boxes are answered with Y.

The lamp is turned on for 30 s and then the program returns to monitor light level. The controller

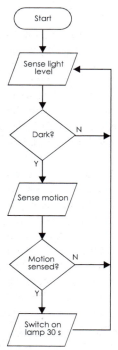

FIGURE 38.6   Security lamp control system.

cycles through the program indefinitely, so this flowchart has no END box.

5) Logical OR is implemented by two decision boxes connected so that action is taken if either of them yields a Y response. The alarm is sounded if the motion sensor detects a person OR if the pressure pad is trodden on.

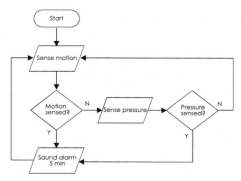

FIGURE 38.7   Security system with two sensors. Either one OR the other can trigger the alarm.

TIMING ROUTINES

6) Delay routine. One way of creating a delay in a program is to make the controller repeat a loop operation a few hundred times. There need be nothing specific for it to do in the loop. The time taken to decrement the loop counter and jump back to the beginning of the loop is multiplied by the number of times the loop is repeated.

Shown in Figure 38.8 is a routine for more extended delays. These are **nested loops**. The loop is repeated m times for every one of the n repetitions of the outer loop. This means that the inner loop is repeated **mn** times altogether. If m and n are both 1000, the inner loop is repeated a million times.

The times obtained with a loop delay routine are not precise. They depend on the rate of the system clock, which is not usually of high precision. Also there may be interrupts (see later) which increase the delay.

7) Timing an operation with a controller generally makes use of a timer register that is incremented at every fetch-execute cycle. With the high speed of the clock and the limitation of the timer register to 1 byte, the overall time period is very short. Periods can be lengthened by hardware dividers (prescalers) in the controller and by using nested program loops.

If a high-precison crystal is used for the system clock, timing can be accurate. However, the machine code or assembler programs for implementing a timing circuit are too complex to be included in this topic. On the other hand, they are extremely simple

FIGURE 38.8 Nested loops provide long delays in a program.

in BASIC or ladder logic and some examples are given later for these languages.

HANDLING DATA

8) **Storing data** in data RAM. When the power is switched off, the values in the controller's registers are lost. This is because the data RAM is volatile — it does not store data permanently.

Most controllers are provided with more permanent RAM in which data can be stored for many years or until it is electrically erased or overwritten with new data. Many of the newer types of controller have flash memory for this purpose.

Data is written into data RAM a byte at a time. The writing routine is given the value to be stored and the address at which to store it. The reading routine is given the address to read and the register to store it in. The routines can be placed in a loop (see 2), 'Repeat an Action', Fig. 38.4) for writing or reading a block of data bytes.

Data may also be transferred serially between the controller and another device. This is done by an on-chip Universal Synchronous Asynchronous Receiver Transmitter (USART). In transmitter mode, a byte placed in a register is automatically coded and transmitted as a series of pulses from one of the controller's output pins. Conversely, when in receiver mode, a series of pulses with the correct format received from an external device is decoded and placed in its receiver register.

INPUT ROUTINES

9) **Registering a keypress** from simple input circuits like those in Figure 36.1 is a matter of repeatedly reading the value of the corresponding bit or input pin. The same technique is used to read the input from sensor interfaces such as that shown in Figure 36.2. This can be done by putting the read command into a loop and testing the result.

In the flowchart below the push-button is connected to the pin identified as pin 3 on the memory map. The interface is like that shown in Figure 36.1a, so the input is normally low but goes high when the button is pressed. The controller circulates around the top loop, continually reading the input, until it returns '1'.

FIGURE 38.9 Waiting for a keypress.

The controller then drops out of the loop to perform the action. After completing this, it returns to the beginning to await the next keypress.

This routine is very simple but it has one problem. Unless the action takes several seconds to perform, the operator may still be holding the key down when the action is finished and the controller has looped back to the beginning. Present-day microelectronic systems operate so fast that they can leave the operator far behind. To prevent this happening when detecting a keypress, we add a second waiting stage to the routine (Figure 38.10).

These small program segments can be used in a number of ways. For instance, in the flowchart in Figure 38.11, another version is included at several points in the program. This does not wait for the keypress, but reads it in passing. Action 5 is performed if the key is *already pressed*, but not performed if the key is not pressed.

The main program cycles through a series of four different actions that are labelled Action 1 to Action 4. At certain stages (two in this example,

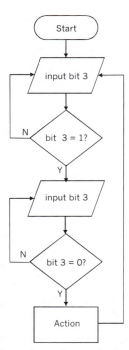

FIGURE 38.10 This routine waits for the key to be pressed, then waits for it to be released before proceeding.

though there could be more) the main program calls up a **subroutine**.

The subroutine is called after Action 1 and performs Action 5 if the key is pressed. The controller is then returned to continue cycling through the other actions. It goes to the subroutine again after Action 3, and there may perform Action 5 again.

As this example shows, after a subroutine is finished, the controller always returns the processor to the point in the main program at which the subroutine was called.

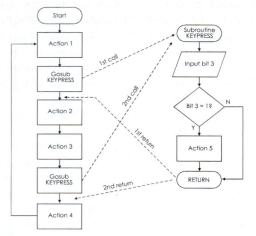

FIGURE 38.11 The main program calls the subroutine to find out if the button is pressed.

Action 5 could include accepting fresh data that modifies Actions 1 to 4. This is a way of allowing the user to change the operating mode of the device by pressing a button at any time.

In a longer program, the subroutine could be called many times. Repeatedly reading a register (or a bit) to determine its current state is called **polling**. An alternative to polling is to use **interrupts**, as explained later.

10) **Counting events,** such as the number of packages passing by on a conveyor belt, are programmed in the flow chart shown in Figure 38.12. This flowchart illustrates two important principles. The first is that all stored values must be **initialised** at the beginning of a program. It is not sufficient to assume that a given bit will be '0' to begin with. It might have been set to '1' by a previously run program. Here, the first thing is to set the counter value to zero.

The sensor might be a photodiode with a beam of light shining on it across the conveyor belt. Its interface is like the circuit shown in Figure 2.5, but with a connection to the 'bit 4' pin of the controller, instead of to the gate of the MOSFET.

Normally, the pin is at 0 V. When an object breaks the beam across the conveyor belt, the voltage at the pin rises close to the supply value, and the pin is at logic 1. While there is no object blocking the light, bit 4 equals 0, and the controller repeatedly reads the bit, waiting for bit 4 to change. When an object breaks the beam, bit 4 changes to 1. The controller drops through the first decision box.

It is not good enough to test the bit only once. We must check that object has passed by and the bit has returned to 0.

The second check is needed because computer programs run very fast! If we do not wait for the bit to clear, the controller might run several times around the loop before the object eventually passes by and bit 4 goes back to 0.

Therefore, we need a second waiting loop, waiting for bit 4 to return to 0. As soon as this has happened it is safe to continue by incrementing *counter*. The next stage is to display the value in the counter byte. Finally, an action, perhaps a 'beep', completes the main program. The controller then loops back, to wait for the next object to arrive.

11) **Input from an analogue sensor** is converted to digital form, either by an external ADC or by an ADC on the controller chip. An external ADC is connected to a port, usually consisting of 8 input/output pins. Data from these pins, though readable individually for use with 1-bit inputs, are in this case read as a single byte. If an on-board ADC is

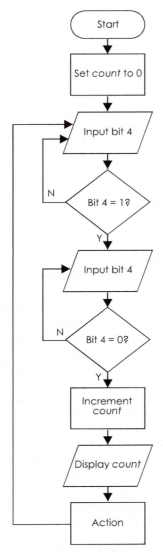

FIGURE 38.12 Counting events requires two checks on the input level.

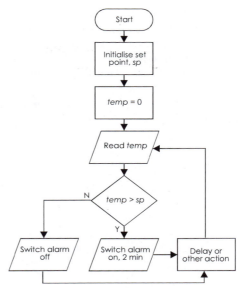

FIGURE 38.13 An over-heating alarm is triggered whenever the temperature exceeds the set point.

used, there is a single pin that accepts an analogue voltage. The voltage is converted and the result appears on a special register, which can be read by the CPU.

The flowchart in Figure 38.13 receives multi-bit input from a thermal sensor. This it compares with a set point value stored in a register. This could have been written into the program, or entered from a keyboard using a separate routine.

At the decision box, the temperature is compared with the set point. The alarm is sounded if the set point is exceeded. Comparison is repeated and the alarm switched off if the temperature falls below the set point.

Note the delay box in the flowchart. This may be a simple time delay (say 1 min) or perhaps a

series of other actions that will take time to occur. This delay prevents the system from oscillating rapidly when close to the set point.

Another technique to avoid oscillations is to program with two levels, a high threshold (HT) and a low threshold (LT). HT can be a few degrees higher than the required temperature and LT a few degrees below it.

The temperature is read and compared with HT. If it is greater than HT the alarm is switched on. If it is lower than LT, the alarm is switched off. If it is between LT and HT there is no change: the alarm is left on if it is already on, or left off if it is already off. This incorporates **hysteresis** into the action. The action is the same as that of the non-inverting Schmitt trgger circuit described in Topic 12.

12) **Contact bounce** is a problem with some types of switch. The program responds so fast that a single key-press is read as many. The hardware solution is shown below.

The software solution is to include a delay in the input routine. This gives time for the input

FIGURE 38.14 The flip-flop changes state almost instantly on first contact and remains in that state until the switch is set the other way.

level to settle at its new value. The flowchart below counts how many times a button is pressed. It leaves the loop after 10 keypresses.

After detecting and registering a press there is a short delay to give the input time to settle.

The program loops back for the next press after checking that the button has been released, followed by another delay for settling.

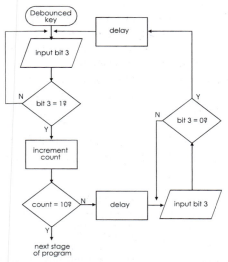

FIGURE 38.15   Inserting delays into a routine is a way to avoid the effect of contact bounce, though it may slow down the action of the program.

13) **Comparator.** Some controllers have an on-board comparator. Two input pins allow the analogue voltage from a sensor to be compared with a reference voltage from a potential divider or a standard voltage source. There may be an on-chip reference.

The output of the comparator (0 or 1) appears as a bit at a specified address in memory. This tells the controller whether the sensor output is greater or less than the set point voltage. The controller is programmed to take appropriate action.

OUTPUT ROUTINES

14) **Flash an LED** once, indefinitely or a given number of times. The LED is controlled by one of the single-bit outputs, or one of the lines of an output port. This can be made low or high to turn the LED off or on.

An LED can be driven directly from the output, as discussed in 'One-bit Output' Topic 36. Alternatively, a high level is used to turn on a MOSFET switch or a BJT switch, as outlined in 'Light Sensitive Alarm', Topic 2. These routines can be used to drive LEDs, lamps, motors, relays and many other electrically powered devices. The routine in Figure 38.16 flashes the LED once.

FIGURE 38.16   This flowchart generates a single flash.

The flowchart above assumes that bit 0 is already configured as an output, so that setting the bit makes its output level change to high. Controller outputs may act either as sources or sinks, so a directly driven LED could be connected through a series resistor to the positive supply line. The bit is made 0 to turn the LED on and 1 to turn it off.

The delay could be a routine like that discussed earlier under 'Timing Routines'. To flash the LED indefinitely, follow the second output box with another delay, then loop back to the beginning.

Flashing a given number of times (Figure 38.17) requires a loop with a loop counter. In this flowchart the controller waits for a keypress before starting to flash. It makes the program simple and shorter if, instead of setting out the delay routine every time it is used, we make it into a subroutine which is called when needed.

15) **Lookup tables** are useful for holding arbitrary values that cannot be calculated from a formula. As an example, in the codes for driving a 7-segment LED (or LCD) the data for each numeral is held in memory for reference by the display routine.

Address (hex)	Data (binary code)
2000H	01111110
2001H	00110000
2002H	01101101
2003H	01111001
2004H	00110011
2005H	01011011
2006H	01011111
2007H	01110000
2008H	01111111
2009H	01111011

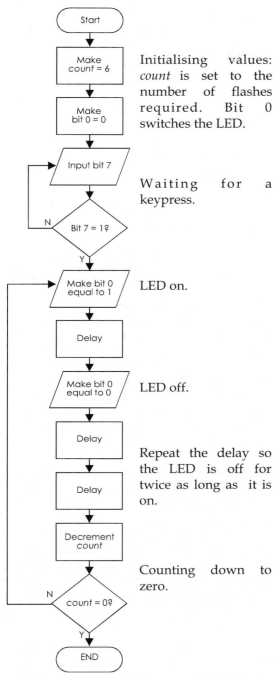

FIGURE 38.17 A routine to flash an LED 6 times after a button is pressed.

In the flowchart below, we are using 7 out of the 8 bits of Port A. Seven of the pins are connected to the LED segments *a* to *g,* the eighth pin being unconnected. The segment is to be lit when the output goes high, so this routine requires a common-cathode display.

Initialisation requires two operations: setting the *count* variable to zero, and switching off all the segments of the display.

The remainder of the program is a loop that repeats indefinitely. Each time round the loop, *count* is incremented by 1. Each time round the route a byte is taken from the look-up table. The address is 200nH, where n is the current value of count. This makes the display show the digits 0 to 9, consecutively.

FIGURE 38.18 Using a look-up table to cycle a 7-segment display through numbers 0 to 9.

The base address of the table on the previous page is 2000H. In other words, the table is stored in memory at addresses 2000H to 2009H.

The codes use all except the first bit of each byte. The allocation of bits to the segments of the display is:

0 a b c d e f g

16) **Wave forms** are easily generated by a controller. A square wave output is produced by a program similar to the indefinite LED flash (see step 14). Using a much shorter delay time produces a square wave at audio frequencies.

A sawtooth wave form is generated by the flowchart shown in Figure 38.19. The output is produced at an 8-bit port, which is connected to a

DAC. Alternatively, we use an on-board DAC and take the analogue signal from its pin.

The program runs in a continuous loop, incrementing *count* each time. So *count* gradually increases from 0 to 255 (8 bits). At the next increment it changes to 0 again, the carry-bit being ignored. It increases from 0 to 255, repeating. The result is a rising voltage from the DAC, dropping instantly to 0, and then repeating. The frequency is set by the length of the delay.

FIGURE 38.19 An easy way to generate a sawtooth waveform.

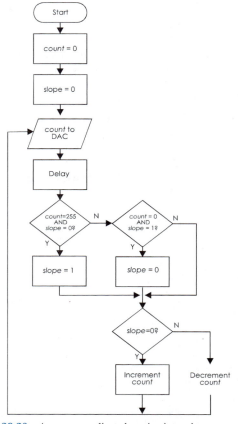

FIGURE 38.20 A more complicated routine is used to generate a sawtooth wave form.

A triangular wave form, in which the voltage ramps both up and down, requires a more complicated flowchart (above right). This uses a single bit, called *slope,* to register whether the output is currently rising (*slope*=0) or falling (*slope*=1).

This flowchart has two decision boxes to detect:

- When *count* is rising AND when it has reached its maximum value (255).
- When *count* is falling AND has reached its minimum value (0).

In practice, each of these decisions based on AND might need to be expanded to two decisions connected as in Figure 38.6. The two AND decisons are connected to provide OR logic, as in Figure 38.7.

The result is that if *count* has reached its maximum rising value OR its minimum falling value, the value of *slope* is changed.

The third decision box, either increments or decrements *count* according to the value of *slope*.

Wave forms that have a more complex shape can usually be generated by using a look-up table. The tabulated values are called up in sequence and sent to the DAC.

17) **Stepper motor** control uses a look-up table which holds the codes for energising the coils (see 'Rotational Speed Controls', Topic 33). The flowchart shown in Figure 38.21 consists of a loop in which the count repeatedly runs from 0 to 3. At each stage the appropriate code is looked up in the table and sent to the coils of the motor (using four transistor switches).

The rate of rotation of the motor depends on the length of the delay. This is a subroutine (not shown here) similar to that shown in Figure 38.8. The instructions to increase or decrease the delay change the values of *n* or *m*.

The circuit has two pushbuttons, 'Faster' and 'Slower'. Each time round the loop, the states of these buttons are read. If either one is pressed, the delay if shortened or lengthened.

18) **Automatic door control** provides an example of program development and illustrates some general points about programming control systems. The doors are driven by a reversible motor that can open or close them.

The sensor in this system is a microwave device that detects persons approaching the door. There

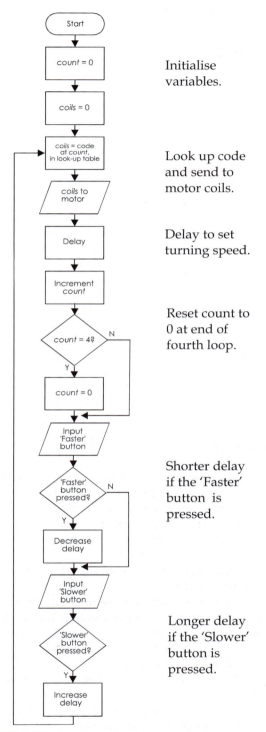

FIGURE 38.21 A routine to control the speed of a stepper motor by using two buttons.

are two outputs to control the motor. One is a relay switched by a transistor switch. This turns the motor off or on. The second output goes to a second relay wired to operate as a reversing switch. In this flowchart (see below) we do not show these outputs separately, but simply state 'Close doors' and 'Open doors'.

After initialising the pins as inputs and outputs and closing the doors for security, the controller waits for person to approach the doors.

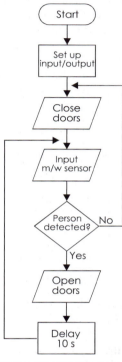

FIGURE 38.22   Controlling the automatic doors of a supermarket.

The controller cycles round the waiting loop until a person approaches. Then the doors are opened. There is a 10 s delay to allow the person to enter, and then the doors are closed. If the person is walking slowly and has not passed through the doorway, the system detects this and keeps the doors open. After the doors are closed, the system waits for the next approaching person.

This flowchart is a good beginning for a program as it covers the essentials, but there are details that need to be thought about. A practical point is that when several people approach the doors in quick succession, the doors do not need opening if they are already open. Another point is that just switching on a motor for a given time does not guarantee that the doors are fully open. We need positive input about whether the doors are fully open or not. This kind of data, about what has (or has not) *actually* happened is a kind of **feedback**.

To obtain this feedback, the system needs two more sensors. A set of microswitches are placed so that one switch closes when the doors are fully open and another switch closes when the doors are fully closed.

We write these into the previous flowchart by including the program segment below in place of the 'Open doors' stage.

The first step is to check if the doors are already open. At the 'Doors shut' box the controller reads the microswitch that indicates this condition. If the doors are not already shut, they are taken to be open and the program skips directly to the end of the open doors routine. If they are already shut, the motor is turned on to open the doors. Then the 'doors open' microswitch is read to find out if the doors have opened as wide as possible. If not, the routine loops around, with the doors motor still running, to check again. When the 'doors-open' sensor confirms that the doors are open, the motor is switched off.

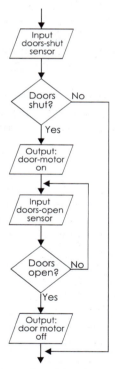

FIGURE 38.23 This expansion of the 'Open doors' routine gives better control of the doors.

A similar routine is needed to expand the 'Close doors' stage.

There are many other ways in which the system can be expanded and improved. For instance, as in many control systems, it is useful to have a manual override button. In this case it would allow the doors to be locked shut when the store is closed. It would also keep the doors closed while the window-cleaner is at work.

INTERRUPTS

A miicroprocessor is always busy handling the program, proceeding from one instruction to the next unless told to stop. Events sometimes occur which need to interrupt this steady progress. An immediate response is essential. Such events include:

- **Arithmetic errors:** a typical example is 'division by zero'. It may happen that a particular term in a calculation is the divisor in an equation. If this term occasionally evaluates to zero, the division can not be performed. This should never happen in a correctly written program, but may sometimes occur during the early stages of development of a new program. This error would be detected inside the controller and causes it to interrupt itself. Instead of going on to the next instruction, it jumps to a special **interrupt service routine** (ISR). If the program is in BASIC, this causes a message to be displayed saying 'Division by zero error'. The microprocessor is then halted and the programmer checks to find why such an error occurred. Acting on reported errors of this kind, the programmer revises the program to eliminate them. We call this **debugging**. Failure to find arithmetic errors may cause the program to 'hang' or to 'crash' when the final version is being run. Debugging is a very essential, and often time consuming, part of programming.
- **Logical errors:** these may have effects as serious as arithmetic errors.
- **Clocked interrupts:** A computer may contain a real-time clock, which can be written to (to set the time and to issue instructions), or read from (to find the date and time). As well as general time-keeping, it can act as an alarm clock, interrupting the controller at any preset time, daily, or on a particular date. It can also be set to interrupt the microprocessor at regular intervals, such as once a second, or once a minute or once an hour.

The microprocessor is programmed to perform a particular action when it is interrupted. In a data logger, for example, it could be programmed to measure and record a temperature at 1-minute intervals, but to continue with other tasks in between. After completing the task that is programmed as an interrupt service routine, the controller returns to the point at which it left its main program. In the data logger, the measurements could be taken so quickly that the user would be unaware that the program had been interrupted.

- **Peripherals:** In a computer system there are several devices that frequently need to interrupt the microprocessor. These devices include the keyboard, the hard disk drive and the printer.

It is possible that a microprocessor receives interrupt signals simultaneously from more than one source. Each device that may interrupt is allocated a priority. This is done by programming or by connecting the devices to different pins with different priorities.

Microcontrollers too may have interrupts, including:

- **Change of input:** The voltage at a particular input has changed from high to low or low to high. This could be the result of opening or closing a switch.
- **Timer overflow.** A timer has reached the end of its set period. Also there may be a built-in watchdog timer (see Extension box 23, Topic 37).
- **USART** has received a new byte.
- **Writing data** to EEROM is completed.

As an alternative to interrupts, a system may rely on **polling**. In some computer systems each input device (which may be something complex such as a keyboard or input port IC) has an **interrupt flag** in one of its registers. In between its main activities, the controller interrogates each device to find out which one or more of them has its interrupt flag set. It then takes action, servicing those with the highest priority first.

Similarly, certain registers in a microcontroller may hold **flag bits**. Each bit is normally 0 but goes to 1 when a given event occurs. The controller is programmed to poll these bits and to act accordingly.

Polling is generally slower than using interrupts because a device or flag bit wait to be interrogated and, by then, it may be too late to avoid a program crash or a loss of data. It is possible to program some controllers to ignore interrupts during certain critical stages in their operations.

Many programmers use interrupts but there are some who prefer polling. They say that interrupts introduce uncertainty into the way the program runs.

DIRECT AND INDIRECT ADDRESSING

Normally a program addresses stored data directly. For example, the controller might be instructed to increment a given value that is stored at a given address. The opcode includes all the essential information — it might say 'Increment the value stored at 037h and place the result in the working register'.

The address is included in the 14-bit opcode. It is the actual address at which the data is stored. This is **direct addressing**.

Because the address is part of the opcode, it can not be altered while the program is running. The instruction will always result in incrementing the value stored at that same address. It can not be changed so as to increment the value stored at any other address. What if we want to increment the values at several different addresses? We use **indirect addressing**.

With indirect addressing the opcode says something slightly different. It might say 'Look in the file select register (FSR) and there you will find an address. Go to that address, increment the value stored there, and place the result in the working register'.

Note that this instruction does not include the address to be incremented, but tells the controller to look in the FSR to find it. The content of the FSR can be changed while the program is running. After each change the FSR sends the controller to a different register to perform its action.

Indirect addressing is particularly useful for operating on a block of stored data. The flowchart below shows how.

Given the first and last addresses in the block, the routine clears them all to 00h, one at a time. A loop is by far the shortest way to program a repeated operation.

FIGURE 38.24 Indirect addressing simplifies operations on blocks of data.

Other block operations can be programmed like this, such as adding, say, 20 to each stored value. Or a block of data can be searched to find a register or registers holding a given value and to perform a given operation on it. For example it could find all registers that hold more than 100 and subtract 30 from them.

THE STACK

In most controllers a small area of RAM is set aside for the stack. This is used for the temporary storage of data. It works in a 'last-in-first-out' manner, like the plate dispenser sometimes seen in canteens. The stack pointer holds the address of the top of the stack. If a byte of data held in the working register is 'pushed' on to the top of the stack, its value is transferred to the RAM location immediately above the current 'top of stack'.

The stack pointer is incremented to point to the new top-of-stack.

The data is held there until it is removed by the 'pop' operation. It is then transferred back to the working register and the stack pointer is decremented to indicate the new location that is top-of-stack. Several data bytes can be pushed or popped between the working register and the stack.

The stack is often used for storing an intermediate result in a calculation. It is also used for registering the state of operations when the CPU jumps to a subroutine, or when the controller is responding to an interrupt.

Immediately before it leaves its current place in the main program, the CPU pushes on to the stack such information as the current contents of the working register, the present value of the program counter, and the states of the flags in the status register. It then jumps to the subroutine or to an ISR. When it returns from the subroutine it pops the data back (in the reverse order) so that the CPU recovers the state that it was in before it jumped. It is then ready to continue with the main program.

SUMMARY

The flowcharts in this topic demonstrate the following program features. The numbers in brackets list the step numbers corresponding to the flowcharts that show each feature:

- Initialise values of variables, look-up tables, define pins and ports as inputs or outputs. (Fig. 38.2)
- Program a sequence of actions. (Fig. 38.2)

- Perform elementary arithmetic and logical operations, including counting. (1, 2, 4, 5, 10, 14B)
- Perform timing operations, including creating time delays. (4, 5, 6, 7, 14, 16)
- Accept input from switches and sensors. (almost all)
- Take decisions as a result of input, arithmetic or logical data. (3, 4, 5, 9A, 9B, 10, 11, 12, 14B, 17, 18)
- Create a sequence of outputs to LEDs, lamps, motors and other actuators, using look-up tables. (15−18)
- Use polling to detect external events. (4, 5, 9, 10, 14B, 17, 18)
- Call on subroutines. (9C)

Most programs incorporate loops, of which there are several kinds:

- Indefinitely repeating loops. (almost all)
- Loops that run for a predetermined number of times. (2, 6, 14B)
- Loops that run until a given condition is true. (almost all)
- Nested loops. (6, 10, 18B)

Where a number is qualified by 'A', 'B' or 'C' it indicates that the feature is present in either the first, second or third flowchart.

The topic also discussed the importance of interrupts, direct and indirect addressing, and stack operations.

ACTIVITIES — PROGRAMMING

Look for ways of extending and improving some of the flowcharts in this topic. Examples of changes you might make are:

Flowchart (3): add a routine to allow a set point to be keyed in.

Flowchart (15): modify this to count *down* from 6 to 1, repeating.

Flowchart (17): adapt the routine to allow the direction of rotation to be reversed.

Flowchart (18): extend this to provide for manual override.

Plan programs for the following tasks and draw the flowchart. Outline what input and output hardware would be required, but do not describe circuit details:

- To control a heater in a thermostat program (see the analogue thermostat shown in Figure 33.2).
- To operate a security system for a lock-up garage that has a broken-beam intruder detector outside the main door, a pressure mat inside the side door, and a vibration detector on the main door. The alarm is given by a siren and a flashing beacon.

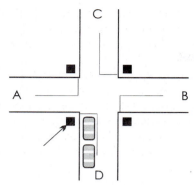

FIGURE 38.25

- A traffic light controller that produces the standard sequence of red, amber and green lights. The lights are at a crossroads (above), but your program need show only the program for the light indicated by the arrow. Then modify the program so that traffic waiting at the indicated light is not kept waiting if there is no approaching traffic on roads A and B.

- Program the following sequence. When a car is waiting at a car park exit barrier, an illuminated sign requests the driver to put 50p in the slot. If this is done the barrier is raised and the car may drive away. The barrier is lowered as soon as the car has gone.

- Design a program to automatically control the points of a model train layout, so that the train circulates in a regular way around three rail loops.

- Design a program for controlling the motion of the conveyor belt at a supermarket checkout.

- Outline a program for timing the laps and displaying the times on a model racing car layout.

Programming Languages

The purpose of this topic is to outline the distinctive features of four different ways of programming a controller. The examples are based on some of the programming flowcharts in Topic 38. This topic does not set out to teach the reader how to program. The only way to learn this is by hands-on experience.

Practice programming on a programming board, using software running on a personal computer. Start with a simple programming task, such as making the controller flash an LED once (Flowchart for step 14). Then extend the routine to flash the LED indefinitely. Then add input from a push-button. By starting with a simple task and gradually adding refinements you will soon learn the essentials of programming a control system.

As an alternative to working with a programming board, there are a number of programs that simulate the action of controllers, allowing you to set up inputs and to read outputs.

PROGRAMMING IN MACHINE CODE

Machine code for the PIC16F690 controller has already been described earlier in Topic 37.

In the example below, we list a machine code program for the flowchart of step 1, which adds together two bytes of data. We assume that the two bytes are already stored in Registers 20h (*num1*) and 21h (*num2*). Their sum is to be placed in Register 22h (*sum*). This is only a segment from a longer program, so it does not start at address 00h. Assume that the sequence of binary instructions is stored in program memory at addresses 70h to 74h. The program is listed in the table at top right.

When programming the controller, it might be necessary to key in the binary code, digit by digit.

Or perhaps the programmer would have a hex keypad for entering the code in hex.

Program Memory Address	Binary Code	Equivalent in Hex
70H	00100000100000	0820h
71H	00000010100010	00A2h
72H	00100000100001	0821h
73H	00011110100010	07A2h
74H	00000001100011	0063h

In either case, the machine code is extremely difficult for a human to read and to check for correctness.

The effect of these instructions is as follows:

At 70H: move the contents of Register 20h to W (the working register, or accumulator).

At 71H: move the contents of W to Register 22h.

At 72H: move the contents of register 21h to W.

At 73H: add Register 22h to W, and place the result in Register 22h.

At 74H: go into sleep mode.

This program demonstrates the 'many simple steps' by which the controller operates.

PROGRAMMING IN ASSEMBLER

Writing programs in machine code is possible but difficult and is very subject to human error. Although using hex makes it easier to enter the instructions in compact form, a microprocessor may have a hundred or more different opcodes in its instruction set. With such a large number it becomes difficult to remember them.

The solution to this problem is to write the program in **assembler** language. Assembler replaces opcodes with **mnemonics**.

Mnemonics are a set of abbreviations usually of three letters, sometimes more, one for each opcode. They are intended to help the human programmer to distinguish the different operations and remember what each operation does.

Example

The instruction 00100000100000 on the instruction bus of a PIC16F690, equivalent to 0820h, tells the controller to move the contents of register 20h to W (the working register). Instead of 0820hH or its long binary equivalent, we use the mnemonic MOVF, which is short for 'MOVe F', where F is a register file. This is represented by the first 6 bits in the instruction.

The mnemonic is followed by one or two **operands**, that in this case tell the controller which file register (F) to use and where to place the result. The seventh bit is the destination bit and the last 7 bits are the register number. The assembler enables the program to use labels to identify registers, which makes it easier to remember what is stored in the register. It also makes it unnecessary to remember the hex address at which the data is stored. In this example the label for register 20h has already been initialised as 'num1'. The full instruction (opcode mnemonic plus operands) is:

MOVF num1,0

The assembler listing for the machine code program opposite is shown below, as it would appear on the computer screen.

In this example, we are using the labels *num1*, *num2* and *sum* for registers 20h, 21h, and 22h respectively.

The example illustrates another helpful feature of assembler. It is possible to add **comments** to the program. These are typed in on the right, after a semicolon. Comments help other people to understand how the program works. They are also very useful to the original programmer too. On coming back to a program a few weeks after it was written, perhaps to correct or extend it, it is surprisingly difficult to recollect how it works. Comments are helpful reminders.

An assembler allows us to run the whole or sections of a program, or to go through the program one line at a time. At every stage it displays the contents of the accumulator or working register, the flags and all the registers, so that we can see exactly what is happening. We can also put a temporary break in a program so that it stops at a given **breakpoint** to allow us to examine the contents of all the registers at that point. When a program is finished and debugged, we can download it into the program ROM of the controller, or save it on to the hard disk.

Programming in Assembler is a slow matter because we have to issue the controller with instructions one at a time. As this example shows, it usually takes several instructions to perform an elementary action, such as adding two numbers together.

```
MOVF        num1, 0          ; move contents of 20h to W

MOVWF       sum              ; move contents of W to 22h

MOVF        num2, 0          ; move contents of 21h to W

ADDWF       sum, 1           ; add contents of W to 22h and
                             ; place result in 22h

SLEEP                        ; go into sleep mode
```

Programming a PIC in Assembler

An assembler program may be a part of a collection, or suite, of programs for use in developing software for a controller. All the programs are linked so that the programmer can pass from one to another and possibly back again as the software is developed. A suite such as this is often called an **integrated development environment**, or IDE.

A view of an IDE in action is displayed opposite.

(continued)

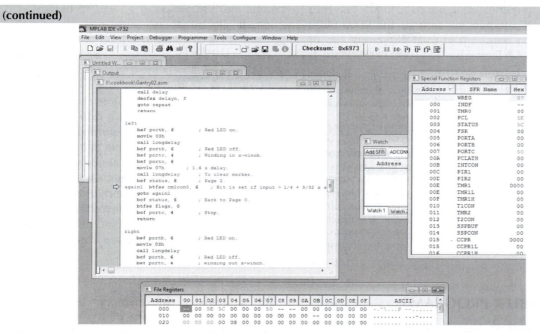

FIGURE 39.1   MPLAB™ is an IDE from Microchip, for programming the PIC series of microcontrollers. Here we are displaying a few of its many windows. The largest window is a text editor where the programmer keys in the program that is being developed. The window on the right shows the contents of the Special Function Registers, including WREG (working register), the STATUS register, and the input/output ports.

MPLAB comprises:
- A text editor.
- A debugger, seen in operation above.
- A simulator, which is software running on the PC but behaving like a PIC.
- Programmer software to download the program into the program memory of the PIC. This uses (among others) the PICkit™ 2 programming device (right). It has a socket for the PIC and an area of board where simple input and output circuits are wired up for testing. It also has LEDs, a push-button and a potentiometer for analogue input.

FIGURE 39.2

If we want to add numbers that total more than 255, or that have many more significant figures, or that are negative, or to add three or more numbers together, the routines become even more complicated when written as single steps. In addition to this, there are dozens of routines that may be used several times in a program. It may be possible to save programming time by including these as subroutines that can be called whenever they are required. Even so, writing a moderately complex program in assembler is a major task.

Programming a PIC in BASIC

There are several IDEs available for programming PICs. One of these is Proton™ IDE, from Mechanique, which has all the expected features of an IDE. A typical working screen is seen below. Another similar IDE is Microcode Studio©, also from Mechanique. There are also compilers such as the one used for programming the BASIC Stamp® controller board from Parallax Inc.

Similar software is available for programming in other high-level languages such as C and its variants. C is widely used by professional programmers but is more difficult to learn than BASIC.

FIGURE 39.3   A short test program listed in the editor window of Proton IDE is ready to be compiled and run.

This program is intended to be run with seven of the Stamp's input/output pins (P0 to P6) connected to the anode pins of a 7-segment common cathode LED display. There is a 1 kΩ resistor in series with each segment to limit the current. The program begins by initialising the input/output pins as inputs or outputs. The bit values in the variable DIR (direction) are 0 for input and 1 for output. Here, pins 0 to 6 are all defined as outputs, while the unused pin 7 is defined as an input.

There is no need to initialise the contents of storage memory bytes B0 and B1, as these are automatically set to zero when the controller starts to run.

Next comes a line label, 'LOOP:' This is followed by a specification of a look-up table. The numbers in brackets are the bit patterns for the 7-segment digits. B0 determines which value is selected from the look-up table. B1 is the byte to which the value is copied.

Example:

On count 3, B0 = 3 and B1 takes the value 91. In binary, this is 0101 1011, the code for the numeral '2'.

The value of B1 is placed on the pins. Where there is '1' the pin goes to logic high, turning on the corresponding segment of the display. After a pause to allow the number to be displayed, B0 is incremented. It is tested to see if it has reached 10 or not. If it has not reached 10, the program jumps back to LOOP to display the next numeral. If it has reached 10 there is a jump to the ZERO label, where B0 is made equal to zero. Then the program jumps back to LOOP to begin again.

PROGRAMMING IN LADDER LOGIC

Ladder logic is a special high-level language, used with programmable logic controllers (PLCs). It is particularly suitable for programming control systems. There are several different forms of ladder logic. Here we illustrate ladder logic using the symbols for the Mitsubishi PLCs.

It is important to be clear from the start that processing a ladder logic program is very different from processing a program written in assembler, BASIC or any other high-level language. In these, the controller starts at the beginning of the program, fetching and executing instructions as it moves from address to address in program memory. Occasionally it may meet a branch instruction and skip to some other part of the program. It may jump to a subroutine and then return to the main program. Once it has begun, the path it takes is determined by the instructions it meets on its way.

In a PLC, the memory is divided into three main areas: the input area, the program area, and the output area. The PLC receives inputs from (usually) a large number of sensors. The first stage in its operation is to scan all these inputs and register their current states in the input area.

Next, the controller works its way through the program, basing its action on the states of the inputs *at the time they were scanned*. It decides what states the outputs should take, and stores these in the output area. The outputs themselves do not change while the controller is working through the program.

Finally, it runs through the output area and updates the outputs accordingly. The controller then cycles back to scan the inputs and the process is repeated.

Typically, the 'input-process-output' cycle takes about 0.1 s. This means that the PLC does not make an *immediate* response to a change in its inputs, but the delay is insignificant in most control applications.

Like other microcontroller systems, a PLC is programmed by software running on a special programmer device or on a personal computer. Ladder logic is represented by a distinctive set of symbols so it is more conveneient to use a special PLC programmer, which has keys marked with these symbols. When a program has been entered and debugged it is downloaded into the memory of the PLC.

Ladder logic is derived from **relay logic**. Relays can be wired together to build logic circuits that can control industrial machinery. With the coming of microelectronics, relay systems have been mainly (though not totally) replaced by PLCs. However, the symbols used for writing ladder logic programs are derived from the symbols used in drawing relay systems. The basic symbol for inputs resembles the symbol for a pair of relay contacts. These may be 'normally open' or 'normally closed'. The symbol for

an output is a circle, similar to the symbol for a lamp or a motor.

FIGURE 39.4   Ladder logic symbols for (a) input contacts normally open, (b) input contacts normally closed, and (c) an output device.

The symbols are labelled with a reference number, X for input, Y for output. The labelling system is different in some versions of the language.

The diagrams opposite show how the symbols are connected together to represent one or more logical operations.

FIGURE 39.5   The symbols for an instruction look like part of a circuit diagram.

To understand the meaning of the diagram, think of the vertical line on the left as a positive supply line, and the vertical line on the right as the 0 V line. Then, if the two sets of contacts are closed, current flows through them and through the output device, which is a lamp in this example. The lamp comes on.

The set of normally closed contacts represents a sensor that produces a logic low level in darkness. The logic high does not energise the imaginary relay, which has normally closed contacts. In the light, the output of the sensor is logic high, the imaginary relay is energised and the contacts are open.

The set of normally open contacts in the diagram represent a sensor that detects a moving person. When a person is detected, the sensor produces a high output and the 'relay' contacts close.

Thus, if it is dark AND motion is detected, both sets of contacts are closed. Imaginary current flows through both sets of contacts and the lamp comes on.

Remember, there are no actual sets of contacts, and no currents flow through them. Instead the logic of the software turns on the lamp if the input from the light sensor is 0 AND the input from the motion sensor is 1. This instruction produces AND logic.

Some PLCs are programmed as connected logic gates. In such a PLC we represent the logic of the previous diagram with the standard logic circuit diagram shown in Figure 39.6.

$$Z = \bar{A}.B$$

FIGURE 39.6   This logic gate circuit has the same action as the ladder logic instruction on the left.

The action of this logic is not a satisfactory substitute for programming a security lamp referred to in the flowchart in Figure 38.6. The input from a motion detector is momentary and the lamp would go off after a second or two. Figure 39.7 shows an improvement to the system.

FIGURE 39.7   Once it is triggered, the lamp is latched on.

The additional 'set of contacts' have the same reference number as the lamp. They are normally open but close when the lamp is on. Once the lamp has been turned on by motion detected in darkness, these contacts close. The lamp is latched on. It stays on, even if motion is no longer detected, and even in the light.

This is not a satisfactory circuit, but it does illustrate how ladder logic can obtain the OR operation by connecting sets of contacts in parallel.

To prevent the lamp staying on indefinitely, we can include a manual reset button (see Figure 39.8).

The reset button input is programmed as a set of normally closed contacts. When the button is pressed, the contact opens. Assuming it is dark and no motion

FIGURE 39.8   A reset button unlatches the 'circuit'.

is detected, there is no way 'current' can flow to the lamp, which goes out.

To produce the same action as in the flowchart in Figure 38.6 (step 4), we can use an additional imaginary device, a timer. Several different types of timer are available in ladder logic software and the one we use here is a pulse timer. Its contacts are normally open, but close for a period when triggered. The length of the pulse can be preset. In Figure 39.9, the legend 'k30' sets it for a 30-second pulse.

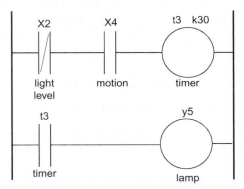

FIGURE 39.9   This is the ladder logic equivalent of the security lamp flowchart.

Instead of the sensors turning on the lamp directly, they trigger the timer. The timer has a pair of normally open contacts, which close when the timer is triggered. The lamp comes on when these contacts close. The contacts stay closed and the lamp stays on for 30 s, even thought the light level and motion inputs may change.

This program has the same action as flowchart 4. We could add more instructions to this program so that the PLC could control an extensive factory security system. The logic diagram will eventually consist of two vertical lines with numerous 'rungs' joining them. This is why this form of programming is called ladder logic.

A counter is another useful item in some ladder logic programs. On the screen it is similar to the drawing Figure 39.10.

FIGURE 39.10 A counter program. Different symbols may be used in different PLCs.

A pulse applied to input 1 resets the counter; its contacts are open and the lamp is off. The counter registers each pulse arriving at input 2. On the tenth pulse, the counter contacts close and the lamp comes on.

After the count has reached 10 the counter contacts stay closed until it is reset.

MULTIPLE CHOICE QUESTIONS

QUESTIONS ON ASSEMBLER

In the assembler routine below, the meanings of the instructions are:

movlw k   Load numeric value k into the working register w (value in hex).
movwf f   Copy value in w to file register f.
incf f, d   Increment register f and place result in destination register (w or f).
decf f, d   Decrement register f and place result in destination register (w or f).
bcf f, b   Clear bit b of file register f.

portc is an 8-bit port set as outputs with an LED driven by bit 3.

Here is the routine:

```
movlw 16h
movwf num ; f label is 'num'
incf num, w
bcf num, 4
decf num, f
movwf portc
```

1 The value in num after the 2nd line is executed is:
   A 0.
   B 16.
   C undecided.
   D 16h.
2 The value in w after the 2nd line is executed is:
   A 16h.
   B 0.
   C 17h.
   D undecided.
3 The value in w after the 3rd line is executed is:
   A 16h.
   B 0.
   C 17h.
   D undecided.
4 The value in num at the end of the routine is:
   A 05h.
   B 16h.
   C 0.
   D 06h.
5 At the end of the routine the LED is:
   A On.
   B Flashing.
   C Not known.
   D Off.
6 After the end of the routine the instruction to turn on an LED connected to portc, bit 4 without affecting the first LED is:

   A bcf portc, 4.
   B incf num, f.
   C bsf portc 17h.
   D bsf portc, 4.

QUESTIONS ON BASIC

In the following PicBASIC routine, the meanings of the instructions are:

Poke address, value   Write value into address. Value is in hex ($) and address is a label.
Pause period   Delay, milliseconds.
Goto   Go to labelled line.

Two LEDs are driven by the outputs at Port C, bits 4 and 5.
Here is the routine:

```
lediode;
   Poke Portc, $20
   Pause 1000
   Poke Portc, $0
   Pause 500
   Goto lediode
```

7 When the routine is running, the LED at bit 4 is:
  A always off.
  B glowing dimly.
  C always on.
  D flashing on and off.

8 When the program is running, the LED at bit 5 is:
  A flashing alternately with the other LED.
  B always off.
  C glowing dimly.
  D flashing.

9 The program stops running:
  A after 1500 ms.
  B until the power is switched off.
  C after the LED has flashed on and off.
  D after 500 ms.

10 To program the LEDs to flash on and off together, change $20 in the second line to:
  A $30.
  B $33.
  C $40.
  D $60.

11 The number of times the LED flashes in a minute is:
  A 500.
  B 1000.
  C 60.
  D 40.

12 The mark-space ratio of the signal switching the LED is:
  A 1.
  B 1.5.
  C 2.
  D 0.67.

QUESTIONS ON LADDER LOGIC

13 The logical operation performed by this instruction is:

FIGURE 39.11

  A NOT.
  B OR.
  C NAND.
  D Exclusive-NOR.

14 The logical operation performed by the instruction below is:

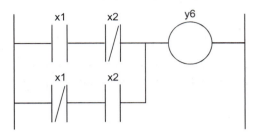

FIGURE 39.12

  A Exclusive-OR.
  B NOR.
  C AND.
  D Latch.

15 In the instruction below, the lamp comes on when:

FIGURE 39.13

  A X2 and X4 are closed.
  B X10 is opened.
  C X10 is closed and the lamp is off.
  D X2 is opened and X4 is closed.

Robotic Systems

A robot is an autonomous machine that is capable of learning new behaviour as the result of experience or training.

Autonomous
Controls itself to a large extent.

Robots are divided into several types:

- **Mobile platform robots**: Most robots are of this type, which includes the android or humanoid robots that look like humans (Fig 40.1), with similar features, but are not necessarily the same size, and which walk on legs. Most mobile robots run on wheels, which are easier than legs to build and program.

FIGURE 40.1 This humanoid robot has an ultrasonic sensor (looking like a pair of eyes, see also Figure 32.6) aimed straight ahead, so that the robot can detect an obstacle in its path before the robot collides with it.

- **Stationary robots**, which include the robot arms that operate on the factory floor performing a wide variety of tasks, such as spray-painting, welding and the assembly of mechanisms from a supply of parts.

ROBOTS IN INDUSTRY

The use of industrial robots has increased enormously in recent years, largely replacing human labour. The reasons for their success are many:

- Precision: a robot operates with a precision of a fraction of a millimetre.
- Repeatability: a robot performs the same task over and over again, without becoming bored, and therefore becoming inaccurate and slow, as a human worker would do.
- A robot can perform tasks that are too arduous, or too dangerous for humans. Fire-fighting, the disposal of noxious chemicals and bomb disposal are examples of this.
- Robots can be built with the power and mechanical strength to lift and manipulate heavy objects.
- They are fast, so are more productive than human workers.
- Robots are controlled by electronic logic, to perform complex tasks, without failure. Their control circuits are based on microprocessors or microcontrollers.
- They are programmed so that they always do the right thing at the right time.
- They are much more dependable than human workers.
- Human workers need a canteen at the factory, for snacks or for their mid-day meal. Robots do not need these facilities.
- Human workers need a place to park a car whilst at work. Robots do not need a car park.

The characteristics listed above apply to other sorts of robot, too, including the humanoid type.

Most robots, except for the robot arms, are able to move from place to place. The humanoid robots walk on a pair of legs. This is difficult to achieve because of the need to keep the robot balanced and upright. If the terrain is rough, the robot may easily stumble and fall. For this reason most robots are either fixed (bolted to a firm base) or run on wheels. Typically, the wheeled robots are three-wheelers, with two wheels powered by a pair of drive motors and a third wheel (a caster, see Figure 40.8) for maintaining balance. It is a fact of solid geometry that a plane surface can be defined by any three points in space, so a three-wheeled robot can be set down on an undulating surface with all three wheels contacting the surface at the same time. Four-wheeled vehicles have the disadvantage that, unless the wheels are individually sprung, it may be that only three of them are in the same plane, in contact with the surface. The vehicle rocks unsteadily as it moves about on an undulating surface. This makes it difficult to steer it accurately and it may lose its balance.

POWER SUPPLIES

Stationary robots, such as robot arms, are easily supplied from a low-voltage power pack driven by the electric mains, but mobile robots *must* be powered by a portable source, such as a battery. Typically the circuits of a robot are built from ICs most of which operate on a 6 V supply. A battery of four AA alkaline cells is used. Rechargeable cells may be used instead by some robot types, since they are cheaper to run than non-rechargeable cells. Solar cells are used for powering robots, such as spacecraft, that operate in remote places.

Unfortunately, most of the small DC motors available for robotics run on 12 V or more, so we need at least two 6 V batteries to power them. The extra weight of the batteries means that the robot is heavier and so, more powerful drive motors are required, making the robot even heavier. This means that we need more powerful, more weighty, batteries and so on.

THE ROBOT AS A SYSTEM

Earlier it was said that a system consists of three parts: input, processing, and output. Robot subsystems have the same three parts but they are differently named. The input to a robot system consists of **sensors** and the outputs are **actuators**. Sensors and actuators are linked through the processing stage, which is a microcontroller.

Normally, a robot comprises many such subsystems, all linked to the same microcontroller, as in Figure 40.2.

ADVANTAGE OF SOFTWARE

The processing in a robotic system is done by a special IC, a microcontroller. Instead of this arrangement, we could devise a robot-like machine consisting solely of electronic sub-circuits such as amplifiers, comparators, and transistor switches.

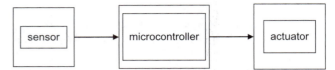

FIGURE 40.2   System diagram of a robot. There may be several sensors, and the robot may receive input from a computer program as well as from sensors.

These sub-circuits would be permanently connected together into one large system - a *hardwired* system. Its behaviour would depend on how the component subsystems were connected to each other, the essential point being that, once they have been connected, the behaviour of the system cannot be changed without disconnecting some of the sub-circuits and reconnecting them to make a different system. The advantage of the microcontroller is that its behaviour can be programmed by *software*. You can change the way the system operates just by altering a few lines in the program. In the more sophisticated robots, the program can be altered by the robot itself while the program is running.

ROBOT SENSORS

Robot sensors usually consist of a sensory component, such as an LDR or a photodiode, with a number of other components, such as resistors and transistors, to interface the sensory component to the microcontroller.

LIGHT SENSOR

This is usually a light dependent resistor or a photodiode. The output from the sensor sub-system is a voltage proportional to the light intensity. This analogue output is compared, using a built-in comparator circuit in the microcontroller, with a standard voltage generated in the microcontroller.

TOUCH SENSITIVE SENSOR

This is usually a microswitch, which clicks on or off, producing a digital output. Touch sensors can be used as proximity detectors, being actuated when the robot runs against an object in its path.

The touch sensor is used like this. The microcontroller is programmed to run forward and repeatedly monitor (poll) the state of the touch sensor. When the sensor is touched the robot stops, runs backward a short distance, spins through a small angle slightly to the left or right, and finally heads off in a new direction. A robot programmed with this

FIGURE 40.3 This robot has two light sensor units, (infra-red photo-diodes) directed downward on to the path ahead. Each sensor also has an LED. Light from this is reflected back from the surface on which the robot is running and is received by the sensor. The amount of light received depends on the colour of the surface. In our experiments the surface was dark brown (hardboard), with strips of light-coloured masking tape stuck on it (see Figure 40.4). The strip makes a pathway for the robot to follow. Normally, the robot runs forward, straddling the path, with the sensors on either side of the path. Both receive a small amount of light reflected from the hardboard. If the path curves to the left, or the sensor swerves to the right, the left-hand sensor will begin to receive light reflected from the masking tape. This increase in brightness is detected and a signal goes to the microcontroller. This is programmed to turn the right motor off, when the left sensor is receiving more light than the right. It leaves the left motor running. The effect of this is to steer the robot to the right, until the sensors are receiving equal amounts of light again. The reverse happens if the path curves to the right or the robot swerves to the left. This has the result of making the robot follow the pathway.

FIGURE 40.5 This robot has an upwardly-directed light sensor, which is used to detect shadows falling on the robot. It is used to give the robot the behaviour of a housefly. If the light falling on the fly is more-or-less constant, it ignores the small changes. But if there is a **sudden fall** in light intensity, the robot dashes away at high speed. It is responding in the same way as a housefly, ignoring small changes in light intensity but, when you try to swat it with your hand, responds to the sudden fall in light intensity as your hand nears the sensor.

FIGURE 40.4 The humanoid robot has downwardly-directed light sensors on its feet. Its program gives it the behaviour of a path-follower, as described for Figure 40.3. Here we see it walking along a pathway of masking tape stuck onto hardboard. This pattern of tape forms a maze, which the robot is programmed to solve. Once it has solved the maze, it always remembers the correct path and runs the maze without making any mistakes.

FIGURE 40.6 The robot has a beam (striped ends) hung from two upright rods. It is forced backward when the robot runs into an object and presses against the touch sensor unit, causing it output to change to logical high.

obstacle-avoidance program runs around the room indefinitely, changing direction whenever it encounters an obstacle in its path.

Touch switches are often used in robotics as **limit switches**. It is not enough for the controller to switch on the motor to extend the arm. It needs to be told when to switch the motor off. A microswitch is mounted on the arm so that it closes when the arm is just fully extended. When that happens the microcontroller switches off the motor and the arm is stopped in the fully extended position.

> The robots illustrated in this topic were constructed and programmed using a LEGO® Mindstorms® kit.

MICROPHONE

A microphone can provide input to a robot so that it can be programmed to respond to sound. For example, it can count hand-claps and respond to the number of handclaps detected. This can be the basis if a steering program in which a single one clap means 'start', two claps mean 'stop', three claps mean 'turn right' and so on.

MAGNETIC FIELD SENSOR

Sensors are available that detect the magnetic field of the Earth (or other magnetic field) and that provide a numerical output indicating the direction in which they are pointing. These can be used to help the robot navigate. The robot in the photograph above has a mast, at the top which the magnetic field sensor is mounted (not visible in photo). The reason for this is that the magnetic fields produced by the robot's drive motors may interfere with the action of detecting the Earth's magnetic field.

FIGURE 40.8   The two drive motors and the castor can be seen in this rear view of a mobile robot.

ROBOT ACTUATORS

Motors

A robot uses actuators to perform actions. In the mobile robots, the most important actuators are the drive motors. The three-wheeled robots have two drive wheels located at the front. The photograph shows the third wheel, which is a caster. Normally, the drive wheels run at the same speed so the robot runs forward or backward in a straight line. If the left wheel is made to turn faster than the right wheel, the robot pursues a course turning to the right. If the right wheel is made to turn faster, the robot steers to the left. If the left wheel turns forward and the right wheel turns backward, the robot spins to the right. If the right wheel turns forward and the left wheel turns backward the robot spins in the opposite direction.

The motors are controlled by an arrangement of transistors known as an H-Bridge. This consists of four MOSFET transistors controlled by two MOSFET inverter gates. Motors are also used in robots to move the arms.

Loudspeaker

A robot can respond to events by making a noise. There are several different types of buzzer or siren that can be switched digitally, or an oscillator may be used to generate a tone of the required frequency. The LEGO Mindstorms robots have a selection of sound files which can be played when required. These include frequently-used words and phrases, such as 'well done', 'goodbye', 'left', 'right', and the

FIGURE 40.7   This photograph shows the LEGO Mindstorms NXT 'brick', which contains the microprocessors and other circuitry of the robot. The upper surface has three control buttons, which provide input to the system. It also has an LCD display for output. At the rear end of the brick (right) you can see the sockets from which current, varying in strength and direction is supplied to the motors. The sockets for connections to the sensors are at the other end of the brick.

numerals 'one' 'two' 'three' and so on, as well as a number of useful sound effects, such as the sound of applause.

DISPLAYS

The simplest form of display is an LED. These can be driven directly from an output pin of the micro-controller, so there are no interfacing considerations. A robot can also be programmed to drive an LCD display, though this is more complicated than driving just a few separate LEDs.

On the Web

Collect information about three different robots. For each one try to find out what it does, how does this, its applications in industry, in robotic research, or in other fields of activity, how it is programmed, who uses such a robot. Write a complete report on the three robots, illustrated By photographs and drawings.

Before writing a program draw a flow-chart. Figure 40.9 is for the obstacle-avoidance program.

Things to do

Take some of the flow-charts from Topic 38 and convert them into assembler language.

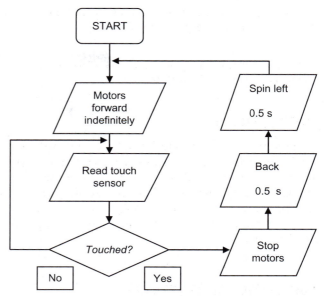

FIGURE 40.9 This robot has a light sensor directed upward. It detects the shadow caused by an object approaching from above, such as a hand about to pick up the robot. In the *Housefly* program, the robot remains stationary, until a hand is brought down, with the intention of picking it up. Then the robot responds by moving away at high speed. With a more complicated program, the robot can be made to respond only if there is a *rapid* change in light intensity, and to ignore a slow change. If the hand approaches slowly and stealthily it is possible to pick up the robot before it escapes. The situation is similar to swatting a fly before it is able to escape.

Neural Networks

Neural networks may be used for controlling or analysing highly complex systems. A neural network might be used for controlling a power station, for example, for landing a lunar module, or for driving a car. Other neural networks are concerned with recognising patterns, as described later.

Animals too have neural networks. They reach their highest state of development in the human brain. A neural network, whether animal or electronic, consists of an assembly of **neurons**. As a very simple example of an electronic neuron, consider the op amp circuit shown in Fig. 41.1.

FIGURE 41.1 An op amp inverting amplifier with weighted inputs is an electronic model of a neuron. The Schmitt trigger inverter ensures that the output is either high or low.

The essence of this circuit is that it:

- accepts **inputs** from a number of sources,
- **processes** them by weighting, summing, and amplifying them, and
- sends **output** to several other circuits.

The weighting is the means by which the behaviour of the neuron is modified. To make the neuron respond more strongly to a given input, the resistor of that input is reduced in value. Then larger current flows to the op amp from that input, increasing its influence on the output of the op amp. To allow this to be done, the input resistors are variable.

In the brain, behaviour in response to various inputs depends on previous experience. In other words, the brain can *learn*. In particular, it usually learns to respond in a way that is beneficial to the animal. To enable the op amp neuron to learn, it must be able to adjust its own weightings. Each neuron needs a small memory in which it stores the results of previous settings of its input resistors. It needs a way of adjusting its input resistors to modify its behaviour in the right directions.

The idea of weighting inputs is fundamental to neural networks. This is because it is not the neurons *themselves* that store information: it is the *connections* between them. As the brain learns a new behaviour, some connections are strengthened, while others are weakened. Using the analogy of the op amp neuron, some resistors are decreased in value while others are increased.

In the previous paragraph we used the word 'analogy'. It is important to remember that the electronic neuron is in many ways *analogous* to an animal neuron, but works in an entirely different manner. In the electronic circuit, the measure of the signal is its *voltage*. In an animal neuron the signal is a series of electrical pulses and the measure of its strength is the *rate of pulsing*. So we must not think of an electronic neural network as a model of a brain.

Analogous

Having a similar action, but operating in a different way or using different structures.

Early **artificial neural networks**, or ANNs, as we call arrays of electronic neurons, were based on op amps. With the development of digital electronics, ANNs are built as VLSI integrated circuits. They have at least the complexity as a Pentium or similar microprocessor and are known as **neurocomputers**. A neurocomputer differs markedly in architecture and operation system from a digital computer.

A digital computer is *programmed* to process input data *serially*. A neurocomputer *accepts data* from the outside world and sends in to be processed by its neural network. It processes the data in *parallel* (all input signals are processed simultaneously). Each time it processes data it may modify the outcome as a result of what it has been taught, or experienced.

The program of a digital computer must necessarily be based on a known set of rules and equations. However, in a neurocomputer, it may be that the values and rules are not completely or precisely known.

For instance, the conditions in the fireball in a furnace are highly complex, and there are many interactions between the combustion of fuel, the flow of air and the generation of heat that are difficult or impossible to predict. In such a case, a neurocomputer would produce more effective control than a digital one. The neurocomputer does not need to know the rules; sometimes there may not be any. All it has to do is to link a required set of outputs to a given set of inputs. It varies its weightings until it achieves as close a match as possible between what the outputs are supposed to be and what they actually are.

MATHEMATICAL MODELS

The action of an op amp and its weighted inputs is essentially mathematical. Mathematical models can be programmed and run on a digital computer. This means that ANNs can be simulated effectively on an ordinary personal computer.

There are several different forms of ANN, many of these still being at the research stage. Their capabilities are being investigated. A form that is often used is the **feedforward network**, shown in Fig. 41.2. Data is input on the left and fed through three or more levels of neuron to the outputs. Other

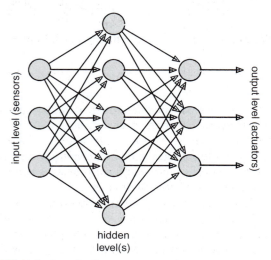

FIGURE 41.2 In a feedforward ANN, each neuron sends signals to the next level in the input-output direction.

forms of ANN have different patterns, some including feedback from the output level to an earlier level.

The input level of an ANN consists of neurons that receive input from sensors, or perhaps data of other kinds. These send signals to each neuron in the **hidden level**. The ANN can decide how many neurons are needed in this level, and also how many hidden levels there need to be. The last hidden level sends signals to each of the neurons of the output level. These is a neuron for each of the possible output signals and these connect to **actuators**.

As a simple example of an ANN, consider a industrial reaction vessel in which a mixture of substances is heated and stirred until a reaction has occurred and the mixture has been converted to a viscous compound.

There are four sensors in the vessel, each of which gives a binary output:

- **Cold**: its output=1 when the temperature is below 40°C.
- **Hot**: its output=1 when the temperature is above 45°C.
- **Viscosity**: its output =1 when the coefficient of viscosity exceeds 100 Pa s.
- **Full**: a level sensor with output=1 when the tank is full to capacity.

The system has four actuators:

- **Heater**: a valve that opens (=1) to allow hot water to circulate in a jacket surrounding the vessel.
- **Cooler**: a valve that opens (=1) to allow cold water to circulate in the jacket.
- **Stirrer**: a motor that drives (=1) a stirrer.
- **Done**: an indicator lamp that lights (=1) when the process is finished.

This system is to be **trained** by **supervision**. We tell the ANN what output is required for every combination of inputs. The required behaviours can be set out as a truth table (see Table 41.1). Because the hot and cold sensors cannot both produce a '1' output at the same time, there are only 12 combinations of input conditions.

The truth table determines what happens at each stage in processing. For example, the stirrer is switched on only when the vessel is full.

The 'done' lamp comes on only when the vessel is full of a viscous mixture between 40°C and 45°C.

The truth table is typed into the computer, and the instruction is given to generate the ANN. The ANN is then put through a series of training runs. To start with, the inputs of each neuron are usually weighted at random. They are varied at each run and, if a given change causes the actual outputs to match more

TABLE 41.1

Row	Inputs				Outputs			
	Cold	Hot	Viscous	Full	Heater	Cooler	Stirrer	Done
0	0	0	0	0	0	0	0	0
1	0	0	0	1	0	0	1	0
2	0	0	1	0	0	0	0	0
3	0	0	1	1	0	0	1	1
4	0	1	0	0	0	0	0	0
5	0	1	0	1	0	1	1	0
6	0	1	1	0	0	0	0	0
7	0	1	1	1	0	0	1	0
8	1	0	0	0	0	0	0	0
9	1	0	0	1	1	0	1	0
10	1	0	1	0	1	0	0	0
11	1	0	1	1	1	0	1	0

closely to the expected outputs, the weightings are changed further in the same direction. Gradually the system discovers the most effective weightings. The mathematical techniques used for changing the weightings is beyond the scope of this account.

In another learning procedure, **reinforcement** training, the network is not told the expected outputs. Instead, its performance is **scored**, depending on how successful it is. It uses this score to modify its weightings and gradually acquires the desired behaviour.

The result is a displayed map of the ANN showing the signal strengths for a given combination of inputs. The map in this example has four input neurons, six neurons in the hidden layer and four in the output layer. After a number of training runs, the four output neurons generate signals as specified in the truth table for each combination of input signals.

This is a simple example, intended to demonstrate what an ANN does. It is obvious that the same result could have been obtained by connecting a few logic gates together. But to connect logic gates correctly we would need to know the logic for every step in the processing. With an ANN we need to know only the initial input and the final output.

PATTERN RECOGNITION

Some of the most successful applications of ANNs have been in the field of pattern recognition. The example already quoted can be taken as a kind of pattern recognition, for the ANN is taught to recognise the input pattern of temperature not over 45°C, temperature not under 40°C, viscosity over 100 Pa s, and

vessel full. When it recognises this pattern it turns on the 'done' sign.

There are other kinds of patterns, such as patterns in financial data. ANNs have been used in various kinds of financial forecasting. Another application is in looking for patterns of spending that indicate fraudulent use of a credit card.

Visual patterns are readily recognised by the human brain. The patterns below are all recognisable as the capital letter 'A'.

We are also good at recognising partial patterns:

neural network

The message is not difficult to read, even though we can see less than half of each letter.

Visual patterns include alphabetic characters, faces, signatures, finger prints and palm prints. There is much use for ANNs in crime prevention and discovery, and in security systems generally.

A digital camera is used for inputting visual patterns. The image is enlarged so as to fill the whole frame. The image is broken down into pixels. In a letter identification system, the input is binary with no shades of grey. A pixel is either 0 (white) or 1 (black). Each pixel provides input to the ANN. In a